AMATEUR ASTRONOMY

The Amateur Astronomer's Library

Under the general editorship of Patrick Moore, F.R.A.S., F.R.S.A.—
a series of books designed for the intelligent amateur, fully illustrated
with linecuts and photographs.

AMATEUR ASTRONOMY

by

PATRICK MOORE, F.R.A.S.
Director of the Armagh Planetarium

New York
W · W · NORTON & COMPANY · INC ·

CONTENTS

5

LIST OF PLATES

7

FOREWORD

WHEN I first wrote the original version of this book, more than a decade ago, amateur astronomy was still in the second phase of its development. Much had been done, and the work of the amateur was recognized and welcomed by professional astronomers all over the world. The Solar System, and in particular the Moon, was still something of an amateur province; until the coming of rocketry, many professionals tended to regard it as somewhat parochial, and this was perfectly logical inasmuch as the Solar System is a very small part of the universe considered as a whole.

Everyone has to make a start sometime, and when writing the book I set out to produce the sort of guide that I would myself have needed in my early days as an observer. I wrote then, and I repeat now, that since I made almost every possible error during my initial observations, I was in some ways qualified to act as a guide to others!

I was particularly glad when the book was well received, and when various new editions came out. But in preparing yet another edition, there was an extra problem to be faced. Amateur astronomy has now moved on into its third phase. Probes have been sent to the Moon and nearer planets, and amateur work has to become more specialized and concentrated. There was no point in simply re-casting those parts of the text which were out of date; it was better to re-write most of the book – and this has been done. I have, of course, retained the original plan, but I hope that the text as it now stands is suited to 1968 instead of the pre-Space Age 1957.

I have had much help, but I must in particular express my thanks to Eric Swenson, of Messrs. W. W. Norton & Co., without whose encouragement this new book would not have appeared.

PATRICK MOORE

Armagh Planetarium, September 1967

AMATEUR ASTRONOMY

Chapter One

ASTRONOMY AS A HOBBY

THE TWENTIETH CENTURY is the Age of Science. Since our grandparents were children, mankind's way of life has changed beyond all recognition, and if we could visit the world of a hundred years ago it would seem almost like being taken to another planet. The idea of a car, an aeroplane or a wireless set would have seemed fantastic to the average mid-Victorian, and even less would he have been prepared for some of the new scientific branches of to-day, such as nuclear physics and radar.

One result of this amazing progress is that science has become specialized. It used to be possible for the amateur to make useful discoveries, while to the normal research worker the possibilities were unlimited. There was always something new "just round the corner", and any apparently trivial experiments could open up new avenues. Such were Becquerel's casual studies of the behaviour of a lump of uranium in a darkened cupboard, which paved the way for the study of radioactivity and a method of treating the dread disease cancer. Now, however, the day of the amateur is largely done. Modern research cannot be carried out without equipment which is so expensive that it can never be assembled by any one man; even theoretical work is beyond the non-specialist.

Astronomy is the one science in which these limitations are not so crippling. The chances of making an important discovery are less than they were at the start of the century, but they still exist. For instance, W. T. Hay—better known as Will Hay, the stage and screen comedian—was the first to see the white spot which appeared on Saturn in 1933, and in the following year a bright "new star" was discovered by another British amateur, J. P. M. Prentice, who happened to be taking an early-morning walk after a period of meteor watching. Moreover, our knowledge of the surface features of our nearest neighbours in space, the Moon and planets, is due largely to the work of amateurs.

It is obvious that there are some branches of astronomy which cannot be tackled by the amateur. The man who builds a 6-inch telescope and sets it up in his back garden can hardly hope to photograph a star-field or a nebula as effectively as an observer using the Palomar 200-inch; nor can he measure the surface temperature of Mars or study the radio waves coming from outer space. But this does not mean that he cannot make himself useful. Professional astronomers, with their great telescopes and complex equipment, have neither the time nor the inclination to make direct studies of objects which are comparatively near at hand. It is true that photographs are taken, but there are times when no photograph can equal sheer visual observations at the eye-end of a telescope.

To drive home this point, it will be useful to give a definite instance of what I mean. In 1955, it was found that the planet Jupiter is a source of "radio waves", or radiation of very long wave-length. This discovery was most interesting, because until then it had been thought that all radio sources lay in the depths of space, well beyond our own Solar System. Research workers wanted to find out whether the waves came from the planet as a whole, or whether they were emitted by a few definite features on the surface. They therefore appealed to the Jupiter Section of the British Astronomical Association, whose members had been making regular observations of the surface features and knew them extremely well. The B.A.A. amateurs suddenly found that their patient labours of past years had become of major importance.

Any serious amateur can do valuable work by making physical observations of the Moon and planets, searching for comets, and studying the fluctuations of variable stars. On the other hand, aimless and haphazard enthusiasm is of no use whatsoever. One has to be systematic, and normally any one observer will confine himself to a particular study, since he will not have time to cover the whole field. Some amateurs concentrate upon variable stars, and show marked annoyance when the Moon drowns faint objects with her light; others spend their astronomical time wholly upon lunar work, and never look at a star except to test their telescopes, while Mars, Jupiter and Saturn all have their followers.

The favourite question asked by the non-enthusiast is:

"What is the immediate use of astronomy? Why do people spend their time watching stars and planets millions of miles away, when there is so much to be done on our own world?"

On the face of it, the question is quite reasonable. It is not immediately obvious why an astronomer should become excited at the appearance of a spot on Saturn, or the flaring-up of a new star. But it must not be forgotten that astronomy is only one branch of science; it has strong links with chemistry, physics and optics, and the stars are vast natural laboratories in which research workers can study matter in unfamiliar states. It is interesting to remember that helium gas, second lightest of the elements, was first found in the Sun. Not until years later was it detected on Earth, and subsequently used to inflate the gas-bags of airships and balloons; its identification was made easier by the fact that its existence was already known.

All timekeeping and navigation is based on astronomy. Greenwich Observatory was originally founded, by order of Charles II, so that a new star catalogue could be drawn up for the use of British seamen. In fact, astronomy is far from being the useless study that so many people imagine. One cannot separate it from other sciences any more than one can separate arithmetic from algebra.

Yet there is another aspect to be considered. In this age of specialization, we are in danger of becoming too concerned with material benefit. What, for instance, is the actual use of a Van Gogh portrait or a Beethoven symphony? The only answer is that a great picture or a great piece of music can give enjoyment to millions of people. And the same is true of astronomy. No painting can equal the sight of the rings of Saturn or the countless stars of the Milky Way; Man can never surpass Nature in her own realm.

Even those who are preoccupied with everyday affairs, and can spare little time for studying the skies, will find astronomy well worth while. If the mildly enthusiastic amateur has no ambition to build or buy a telescope large enough for him to do useful work, he can still give himself hours of pleasure by observing for his own amusement, and as he learns he will find the horizon opening out before him. What does it matter if he never discovers a comet or solves the riddle of the Martian canals? Few people who learn the piano in their spare time have any delusions that they will end by playing as brilliantly as Paderewski.

However, one cannot draw the best out of astronomy without taking some trouble. The night sky becomes far more attractive once the Great Bear, the Dragon and all the other constellations can be recognized on sight, while a planet grows in fascination as its true nature becomes known. The aim of my present book is to explain the basic facts as clearly and as simply as possible, as well as indicating some lines of work which can be undertaken by the amateur who wants to make himself useful to others. It is, in fact, an attempt to answer the second oft-asked question: "If I want to make a hobby of astronomy, how do I go about it?"

Chapter Two

THE UNFOLDING UNIVERSE

A SUBJECT CAN ALWAYS be better understood if something is known about its history. Though we no longer worship our "honourable ancestors", it is a distinct help to look back through time in order to see how knowledge has been built up through the centuries. This is particularly true with astronomy, which is the oldest science in the world—so old, indeed, that we do not know when it began.

Most people of to-day have at least some knowledge of the universe in which we live. The Earth is a globe nearly 8,000 miles in diameter, and is one of nine "planets" revolving round the Sun. The best way of describing the difference between a planet and a star is to say that the Earth is a typical planet, while the Sun is a typical star.

Five planets—Mercury, Venus, Mars, Jupiter and Saturn—were known to the ancients, while three more have been discovered in modern times. Jupiter is the largest of them, and its vast globe could hold more than a thousand bodies the size of the Earth, but even Jupiter is tiny compared with the Sun. The stars of the night-sky are themselves suns, many of them far larger and more brilliant than our own, and appearing small and faint only because they are so far away. On the other hand, the Moon shines more brilliantly than any other body in our skies apart from the Sun. This importance is not real; the Moon is a most insignificant body, and has no light of its own. It has a diameter only one-quarter of that of the Earth, and is by far the closest non-artificial object in the heavens.

The whole celestial vault seems to revolve round the Earth once a day. This apparent motion is due, of course, to the fact that the Earth is spinning on its axis from west to east. Of all the natural bodies in the sky, only the Moon has a real movement round the Earth.

We are used to taking these facts for granted, but at the start of human history it was believed that the Earth was flat and stationary. The Sun and Moon were worshipped as gods, while the appearance of something unusual in the heavens was taken as a sign of divine displeasure.

It is usually said that the first astronomers were the Chaldæans, the Egyptians and the Chinese, but this is only partially correct. It is true that these ancient folk divided the fixed stars into groups or "constellations", and also recorded planets, comets and eclipses, but they had no real understanding of the nature of the universe or even of the Earth itself, so that they were hardly "astronomers" in the full sense of the word.

The story begins in about 3000 B.C., when the 365-day year was first adopted in Egypt and in China. This, too, was the approximate date of the building of that remarkable structure known as the Great Pyramid of Cheops. The Pyramid is still one of the main tourist attractions of Egypt; Cheops himself, a harsh and determined ruler, spent so much money upon it that he ruined his country, and even now we are not certain why he regarded the Pyramid as so important. From an astronomical point of view, it is interesting because its main passage is oriented with the north pole of the sky.

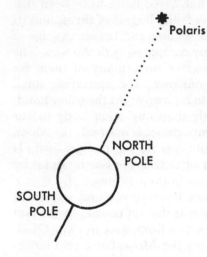

Fig. 1. The axis of the Earth.

The Earth's axis of rotation is inclined at an angle of $23\frac{1}{2}$ degrees, and points northwards to the celestial pole (Fig. 1). To-day the pole is marked approximately by a bright star known as Polaris, familiar to every navigator because it seems to remain almost stationary while the other celestial bodies revolve round it. In Cheops' time, however, the polar point was in a different position, close to a much fainter star, Thuban in the constellation of the Dragon. The reason for this change is that the Earth is "wobbling" slightly, like a top that is about to fall, and the direction of the axis is describing a circle in the sky. The wobbling is very slow, but the shift of the pole has become appreciable since the Pyramid was built 5,000 years ago.

Egypt is still regarded as the land of mystery. As a matter of fact, most of the mysteries of Ancient Egypt were deliberately created by the priests, who were the most learned of their race and who realized that the best way of controlling the common people was to keep them in ignorance. Even the priests had marked limitations, and although they excelled in the art of making exact measurements, and in land survey, they never found out that the Earth is a globe. They believed the world to be rectangular, with Egypt in the middle and deserts and seas all round.

Chinese astronomy was no more advanced. Records of comets and eclipses have come down to us, but some of the ideas held in those times seem strange to-day. One famous story about an eclipse will show what is meant.

The Moon revolves round the Earth once a month, while the Earth revolves round the Sun once a year. The Moon is much smaller than the Sun, but it is also much closer, so that in our skies the two look almost exactly the same size. When the Sun, Moon and Earth move into a straight line, with the Moon in the middle, we see what is known as a solar eclipse; the dark, non-luminous body of the Moon blots out the Sun, and for a few minutes "day is turned into night". If the Moon covers the Sun completely, the eclipse is total.

The Chinese knew about eclipses, and even worked out how to predict them, but they had no idea that the Moon is responsible. They thought that the Sun was in danger of being swallowed by a hungry dragon, and they therefore made it their custom to beat gongs and pans as loudly as possible, hoping that the noise would scare the dragon away. (It always did!) In 2136 B.C. the Court Astronomers, Hsi and Ho, failed to give warning that an eclipse was due, and in consequence no preparations could be made. The luckless two were held to have imperilled the whole world by their neglect of duty, and were executed. The story may or may not be true!

Astronomy in its true form began with the Greeks, who not only made observations but who also tried to explain them. The first of the great philosophers was Thales of Miletus, who was born in 624 B.C.; the last was Ptolemy of Alexandria, and with

19

his death, in or about A.D. 180, the classical period of science comes to an end. During the intervening eight centuries, human thought made remarkable progress.

Thales himself may have been the first to realize that the Earth is a globe, but unfortunately all his original writings have been lost. The first definite arguments against the old idea of a flat Earth are given by Aristotle, who was born in 384 B.C. and died in 322. Aristotle was one of the most brilliant men of the ancient world, and his reasoning shows the Greek mind at its best.

As Aristotle points out, the stars appear to alter in altitude above the horizon according to the latitude of the observer. Polaris appears to remain fairly high in the sky as seen from Greece, because Greece is well north of the terrestial equator; from Egypt, Polaris is lower; from southern latitudes it cannot be seen at all, since it never rises above the horizon. On the other hand Canopus, a brilliant star in the southern part of the sky, can be seen from Egypt but not from Greece. This is just what would be expected on the theory of a round Earth, but such behaviour cannot possibly be explained if we suppose the Earth to be flat. Aristotle also noticed that during a lunar eclipse, when the Earth's shadow falls across the Moon, the edge of the shadow appears curved—indicating that the surface of the Earth must also be curved.

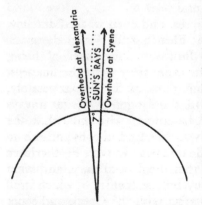

Fig. 2. Eratosthenes' method of measuring the circumference of the Earth.

The next step was taken by Eratosthenes of Cyrene, who succeeded in measuring the length of the Earth's circumference. His method was most ingenious, and proved to be remarkably accurate. Eratosthenes was in charge of a great scientific library at Alexandria, in Egypt, and from one of the books available to him he learned that at the time of the summer solstice, the "longest day" in northern latitudes, the Sun was vertically overhead at noon as seen from the town of

20

Syene (the modern Assouan), some distance up the Nile. At Alexandria, however, the Sun was at this moment 7 degrees away from the overhead point, as is shown in Fig. 2. A full circle contains 360 degrees, and 7 is about 1/50 of 360, so that if the Earth is spherical its circumference must be 50 times the distance from Alexandria to Syene. Eratosthenes may have arrived at the final figure of 24,850 miles, which is only fifty miles too small.*

If the Greeks had taken one step more, and placed the Sun in the centre of the planetary system, the progress of astronomy would have been rapid. Some of the philosophers tried to do so; but unfortunately Aristotle held the Earth to be the centre of the universe, and Aristotle's authority was so great that few people dared to question it. Moreover, the decentralization of the Earth would have meant a change in the laws of "physics", since Aristotle's idea of "things seeking their natural place" would have been much disturbed.

Most of our knowledge of Greek astronomy is due to Claudius Ptolemæus (Ptolemy), who wrote a great book known generally by its Arab title of the *Almagest*. In it, he sums up the ideas of the great philosophers who had lived before him; and the theory that the Earth lies at rest in the centre of the universe is therefore called the "Ptolemaic", though as a matter of fact Ptolemy himself was not directly responsible for it.

On the Ptolemaic theory, all the celestial bodies move round the Earth. Closest to us is the Moon; then come Mercury, Venus, the Sun, Mars, Jupiter, Saturn and finally the stars. Ptolemy maintained that since the circle is the "perfect" form, and nothing short of perfection can be allowed in the heavens, all these bodies must move in circular paths. Unfortunately, the planets have their own ways of behaving. Ptolemy was an excellent mathematician, and he knew quite well that the planetary motions cannot be explained on the hypothesis of uniform circular motion round a central Earth. He therefore worked out a complex system according to which each planet moved in a small circle or "epicycle", the centre of which itself moved round the Earth in a perfect circle. As more and more irregularities came to light, more and more epicycles had to be

---

* There is some doubt as to whether Eratosthenes' estimate was accurate to within a few tens of miles, but at least his results were not wildly in error.

introduced, until the whole system became hopelessly artificial and cumbersome.

Hipparchus, who had lived some two centuries before Ptolemy, had drawn up a detailed and accurate star catalogue. The original has been lost, but fortunately Ptolemy reproduced it in his *Almagest*, so that most of the work has come down to us. Hipparchus was also the inventor of an entirely new branch of mathematics, known to us as trigonometry.

When the power of Greece crumbled away, astronomical progress came to an abrupt halt. The great library at Alexandria was looted and burned in A.D. 640, by order of the Arab caliph Omar, though in fact most of the books may have been scattered earlier; in any case, the loss of the Library books was irreparable, and scholars have never ceased to regret it. For over a thousand years very little was done. When interest in the skies did return, it came—ironically enough—by way of astrology.

Even to-day, there are still some people who do not know the difference between astrology and astronomy. Actually, the two are utterly different. Astronomy is an exact science; astrology is a relic of the past, and there is no scientific basis for it, though in some countries (notably India) it still has a considerable following.

The best way to define astrology is to say that it is the superstition of the stars. Each celestial body is supposed to have a definite influence upon the character and destiny of each human being, and by casting a horoscope, which is basically a chart of the positions of the planets at the time of the subject's birth, an astrologer claims to be able to foretell the destiny of the person for whom the horoscope is cast. There may have been some excuse for this sort of thing in the Dark Ages, but there is none to-day. The best that can be said of astrology is that it is fairly harmless so long as it is confined to circus tents and the less serious columns of the Sunday newspapers.

However, mediæval astrology did at least lead to a revival of true astronomy. The Arabs led the way, and presently interest spread to Europe. Star catalogues were improved, and the movements of the Moon and planets were re-examined. There were even observatories; very different from the domed buildings of to-day, but observatories none the less.

22

Astronomy was still crippled by the blind faith in Ptolemy's system. So long as men refused to believe that the Earth could be in motion, no real progress could be made. The situation was not improved by the attitude of the Church, which in those times was all-powerful. Any criticism of Aristotle was regarded as heresy. Since the usual fate of a heretic was to be burned at the stake, it was clearly unwise to be too candid.

The first serious signs of the approaching struggle came in 1546, with the publication of *De Revolutionibus Orbium Cælestium* (Concerning the Revolutions of the Heavenly Bodies) by a Polish canon, Nicolas Copernicus. Copernicus was a clear thinker, as well as being a skilful mathematician, and at a fairly early stage in his career he saw so many weak links in the Ptolemaic system that he felt bound to abandon it. It seemed unreasonable to suppose that the stars could circle the Earth once a day. In his own words, "Why should we hesitate to grant the Earth a motion natural and corresponding to its spherical form? And why are we not willing to acknowledge that the *appearance* of a daily rotation belongs to the heavens, its *actuality* to the Earth? The relation is similar to that of which Virgil's Æneas said, 'We sail out of the harbour, and the countries and cities recede.' "

Copernicus' next step was even bolder. He saw that the movements of the Sun, Moon and planets could not be explained by the old system even when all Ptolemy's circles and epicycles had been allowed for, and so he rejected the whole theory. He placed the Sun in the centre of the system, and reduced the status of the Earth to that of a perfectly ordinary planet.

Copernicus was wise enough to be cautious. He knew that he was certain to be accused of heresy, and though his book was probably complete by 1530 he refused to publish it until the year of his death. As he had foreseen, the Church was openly hostile. Bitter arguments raged throughout the next half-century, and one philosopher, Giordano Bruno, was burned in Rome because he insisted that Copernicus had been right.*

Tycho Brahe, born in Denmark only a few months after

* This was not Bruno's only crime in the eyes of the Church, but it was certainly a serious one.

Copernicus died, was utterly unlike the gentle, learned Polish mathematician. Tycho was a firm believer in astrology, and an equally firm disbeliever in the Copernican system, so that it is ironical to realize that his own work did much to prove the truth of the new ideas. He built an observatory on the island of Hven, in the Baltic, and between 1576 and 1596 he made thousands of very accurate observations of the positions of the stars and planets, finally producing a catalogue that was far better than Ptolemy's. Of course, he had no telescopes; but his measuring instruments were the best of their time, and Tycho himself was a magnificent observer.

The story of his life would need a complete book to itself. Tycho is, indeed, one of the most fascinating characters in the history of astronomy. He was proud, imperious and grasping, with a wonderful sense of his own importance; he was also landlord of Hven, and the islanders had little cause to love him. His observatory was even equipped with a prison, while his retinue is said to have included a pet dwarf. Yet despite all his shortcomings, he must rank with the intellectual giants of his age. Nowadays, nothing remains of his great Uraniborg observatory.

When Tycho died, in 1601, he left his observations to his assistant, a young German mathematician named Johann Kepler. After years of careful study, Kepler saw that the movements of the planets could be explained neither by circular motion round the Earth, nor by circular motion round the Sun, so that there was something wrong with Copernicus' system as well as with that of Ptolemy. Finally, he found the answer. The planets do indeed revolve round the Sun, but not in perfect circles. Their paths, or "orbits", are elliptical.

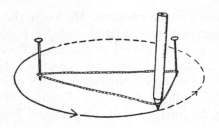

Fig. 3. Method of drawing an ellipse.

One way to draw an ellipse is shown in Fig. 3. Fix two pins in a board, and join them with a thread, leaving a certain amount of slack. Now loop a pencil to the thread, and draw it round the pins, keeping the thread tight. The result will be an

24

ellipse,* and the distance between the two pins or "foci" will be a measure of the eccentricity of the ellipse. If the foci are close together, the eccentricity will be small, and the ellipse very little different from a circle; if the foci are widely separated, the ellipse will be long and narrow.

The five planets known in Kepler's day proved to have paths which were almost circular, *but not quite*. The slight departure from perfect circularity made all the difference, and Tycho's observations fell beautifully into place, like the last pieces of a jig-saw puzzle. The age-old problem had been solved, though the Church authorities continued to oppose the truth for some time longer. Kepler's three Laws of Planetary Motion, the last of which was published in 1618, paved the way for the later work of Sir Isaac Newton.

Kepler's work was not the only important development to enrich the early part of the seventeenth century. In 1608 a spectacle-maker of Middelburg in Holland, Hans Lippersheim, found that by arranging two lenses in a particular way he could obtain magnified views of distant objects. Spectacles had been in use for some time—according to some authorities, they were invented by Roger Bacon—but nobody had hit upon the principle of the telescope until Lippersheim did so, more or less by accident.

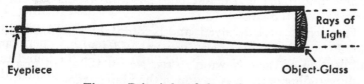

Fig. 4. Principle of the refractor.

A refracting telescope consists basically of two lenses. One, the larger, is the object-glass; its function is to collect the rays of light coming from a distant object, and bunch them together to form an image at the focus (Fig. 4). The image is then magnified by a smaller lens known as an eye-piece. This is more or less the principle used in the naval and hand telescopes of to-day, as well as in ordinary binoculars.

* The method is excellent in theory. In practice, what usually happens is that the pins fall down or the thread breaks. One day, I hope to carry out the whole manœuvre successfully.

The news of the discovery spread across Europe, and came to the ears of Galileo Galilei, Professor of Mathematics at the University of Padua. Galileo was quick to see that the telescope could be put to astronomical use, and "sparing neither trouble nor expense", as he himself wrote, he built an instrument of his own. It was a tiny thing, pitifully feeble compared with a modern pocket telescope, but it helped towards a complete revolution in scientific thought.

Galileo's first telescopic views of the heavens were obtained towards the end of 1609. At once, the universe began to unfold before his eyes. The Moon was covered with dark plains, lofty mountains and giant craters; Venus, the Evening Star of the ancients, presented lunar-type phases, to that it was sometimes crescent, sometimes half and sometimes nearly full; Jupiter was attended by four moons of its own, while the Milky Way proved to be made up of innumerable faint stars.

Galileo had always believed in the new system of the universe, and his telescopic work made him even more certain. Inevitably he found himself in trouble with the Church. It was hard for religious leaders to realize that the Earth is not the most important body in the universe, and Galileo seemed to them to be a dangerous heretic. He was arrested and imprisoned, after which he was brought to trial and forced to "curse and abjure and detest" the false theory that the Earth moves round the Sun.

Few people were deceived, and before the end of the century the Ptolemaic theory had been abandoned for ever. The publication of Isaac Newton's *Principia*, in 1687, led to a real understanding of the way in which the planets move.

Most people have heard the story of Newton and the apple. It is interesting because unlike most stories of similar type, such as Canute and the waves, it is probably true. Apparently Newton was sitting in his garden one day when he saw an apple fall from its branch to the ground, and upon reflection he realized that the force pulling on the apple was the same force as that which keeps the Moon in its path round the Earth. From this he was led on to the idea of "gravitation", upon which the whole of later research has been based. It is fair to say that Kepler found out "how" the planets move; Newton discovered "why" they do so.

Newton also constructed an entirely new type of telescope. As has been shown, Galileo's instrument was a refractor, and used an object-glass to collect its light. Newton came to the conclusion that refractors would never be really satisfactory, and he looked for some way out of the difficulty. Finally he decided to do away with object-glasses altogether, and to collect the light by means of a specially-shaped mirror.

When Newton rejected the refractor as unsatisfactory, he was making one of his rare mistakes. However, the Newtonian "reflector" soon became popular, and has remained so. Mirrors are easier to build than lenses, and even to-day all the world's largest instruments are of the reflecting type.

Astronomy was growing up. So long as observations had to be made with the naked eye alone, little could be learned about the nature of the planets and stars; their movements could be studied, but that was all. As soon as telescopes became available, true observatories made their appearance. Copenhagen and Leyden took the lead; the Paris Observatory was completed in 1671, and Greenwich in 1675.

Greenwich was founded for a special reason. England has always been a seafaring nation, and before the development of reliable clocks the only way in which sailors could fix their position when far out in the ocean, out of sight of land, was to observe the position of the Moon among the stars. This involved the use of a good star catalogue, and the best one available, Tycho's, was still not accurate enough. Charles II therefore ordered that the star places must be "anew observed, examined and corrected for the use of my seamen". A site was selected in the Royal Park at Greenwich, and Sir Christopher Wren, himself a former professor of astronomy, designed the first observatory building. The Rev. John Flamsteed was appointed Astronomer Royal, and in due course the revised star catalogue was completed.

Telescopes continued to be improved. Some of the early instruments were curious indeed; one of them, used by the Dutch observer Christiaan Huygens, was over 200 feet long, so that the object-glass had to be fixed to a mast. But gradually the worst difficulties were overcome, and both refractors and reflectors gained in power and in convenience. Mathematical astronomy made equally rapid strides. The great obstacle had always been the Ptolemaic system, and once that had been swept away the path was clear. The distance between the Earth and the Sun was measured with fair accuracy, and in 1675 the Danish astronomer Ole Rømer even measured the speed of light, which proved to be 186,000 miles per second. Rømer did this, incidentally, by observing the movements of the four bright moons of Jupiter.

But though knowledge of the bodies of the Solar System had improved out of all recognition, little was known about the stars, which were still regarded as mere points of reference. The first serious attack on their problems was made by William Herschel, who is rightly termed the "father of stellar astronomy".

Herschel was born in Hanover in 1738, eleven years after the death of Newton. He came to England, and became organist at the Octagon Chapel in Bath; but his main interest was astronomy, and he built reflecting telescopes which were the best of their age. The largest of Herschel's telescopes, built comparatively late in his career, had a mirror 48 inches in diameter. The mirror still exists, and now hangs on the wall of Flamsteed House in Greenwich, though it has not been used since Herschel's time.

Herschel had his living to earn, and for some years he could not afford to spend all his time in studying astronomy. Then, in 1781, he made a discovery which altered his whole life. One night he was examining some faint stars in the constellation of the Twins, when he came across an object which was certainly not a star. At first he took it for a comet, but as soon as its path was worked out there could no longer be any doubt as to its nature. It was not a comet, but a planet—the world we now call Uranus.

The discovery was quite unexpected. There were five known planets, and these, together with the Sun and Moon, made a grand total of seven. Seven was the magical number of the ancients, and it had therefore been thought that the Solar System must be complete. Herschel became world-famous; he was appointed Court Astronomer to King George III, and henceforth he was able to give up his musical career altogether.

Herschel set himself a tremendous programme. He decided to explore the whole heavens, so that he could form some idea of the way in which the stars were arranged. Until the end of his long life, in 1822, he worked patiently at his task, and his final conclusions have been proved to be extremely accurate.

Naturally, Herschel made numerous discoveries during his sky-sweeps. Many apparently single stars proved to be double, and there were also clusters of stars, as well as faint luminous patches known as "nebulæ", from the Latin word meaning "clouds". Herschel was a most painstaking observer. He catalogued all his discoveries, and when we examine his published papers we can only marvel at the amount of work he managed to do. Since he lived in England for most of his life, he was unable to examine the stars of the far south, which never rise in northern latitudes, and it was fitting that the completion of his sky-sweeps should be accomplished later by his son, Sir John Herschel, who travelled to the Cape of Good Hope specially for the purpose, and remained there for several years.

Another famous observer of this period was Johann Schröter, chief magistrate of the little German town of Lilienthal. Unlike Herschel, Schröter concentrated mainly upon the Moon and planets, and he is the real founder of "selenography", the physical study of the lunar surface. Unfortunately Schröter's observatory, together with all his unpublished work, was destroyed by the invading French armies in 1814, and Schröter himself died two years later.

In the early years of the nineteenth century a German optician, Fraunhofer, began to experiment with glass prisms. Newton had already found that ordinary "white" light is not white at all, but is a blend of all the colours of the rainbow. Fraunhofer realized that this discovery could be turned to good account, and his work led to the development of a new instrument, the astronomical spectroscope.

Just as a telescope collects light, so a spectroscope analyses it. By studying the "spectra" produced, it is possible to find out a great deal about the matter present in the material which is emitting the light. For instance, the spectrum of the Sun shows two dark bands which can be due only to the element sodium, so that we can prove that sodium exists in the Sun.

Spectroscopes can be made by amateurs, but we have to admit that little useful work can be done except with complex and expensive equipment. To the professional astronomer of to-day, the telescope would be of little use without the spectroscope; it is now possible to track down familiar elements in remote stars, and even in other star-systems far away in the depths of space.

In 1838, Friedrich Bessel, Director of the Observatory of Königsberg, returned to the problem of the distances of the stars. By studying the apparent movements of 61 Cygni, a faint object in the constellation of the Swan, he was able to show that it lay at a distance of about 60 million million miles. About the same time a British astronomer, Henderson, measured the distance of the bright southern star Alpha Centauri, and arrived at the reasonably accurate value of twenty million million miles; the real value is about 24 million million miles, so that Henderson underestimated somewhat. Alpha Centauri is a triple star, and the faintest member of the trio remains the nearest known body outside our own Solar System.

Twenty-four million million miles! Our brains are not built to understand such vast distances, and it is clear that the mile is too short to be a convenient unit of length. One might as well try to measure the distance between London and Melbourne in centimetres. Fortunately there is a much better unit available, based upon the speed of light.

Light is known to travel at 186,000 miles per second. A ray from the Sun takes $8\frac{1}{3}$ minutes to reach us, but in the case of Alpha Centauri the time of travel is $4\frac{1}{3}$ years; we see the star not as it is now, but as it was $4\frac{1}{3}$ years ago. Alpha Centauri is therefore said to be $4\frac{1}{3}$ light-years away, while the distance of 61 Cygni is nearly 11 light-years.

Bessel's success gives us an added idea of the real unimportance of the Solar System. Rather than quote strings of figures, it will be more graphic to imagine a scale model. If we begin with making the Sun a 2-foot globe, and putting it on Westminster Bridge, the Earth will become a pea at a distance of 215 feet; Uranus, the outermost of the planets known in Bessel's time, will be represented by a plum $\frac{4}{5}$ of a mile away from our 2-foot Sun. What of the nearest star? We shall not find it in London, or even in England; it will lie some 10,000 miles away, in the frozen wastes of Siberia. We have learned much since the days when the Earth was thought to be the hub of the universe.

Another great event of the last century was the beginning of astronomical photography. In 1845 the first "Daguerreotype" picture of the Sun was taken, followed in 1850 by a good photograph of the Moon. Within fifty years, magnificent photographs of the celestial bodies were being taken not only at the official observatories, but also by amateurs. To-day most of the regular work of the professional astronomer is done with the aid of photography, and sheer visual observation is rare, since in general the photograph is not only more reliable than the eye but also leaves a permanent record. It can also detect objects too faint to be seen by visual means.

Herschel's 48-inch reflector was soon surpassed. In 1845 Lord Rosse, in Ireland, built a 72-inch. It was cumbersome and awkward to use, but it was by far the most powerful instrument then in existence, and Rosse used it to study the clusters and nebulæ which had been pointed out by Herschel. Some of the nebulæ proved to consist entirely of faint stars, though others could not be so resolved. Even more interesting was the fact that some of the starry nebulæ revealed a spiral structure, so that they looked very much like Catherine-wheels.

Alone, the telescope could never decide upon the nature of the irresolvable nebulæ; the spectroscope was able to do so. In 1864 Sir William Huggins examined a faint nebula in the Dragon, and found that it was made up not of stars, but of luminous gas.

It is now known that the nebular objects are of three types. Inside our own star-system, known commonly as the Milky Way but more properly as the Galaxy, we find the normal star-clusters and the gaseous nebulæ, most of them hundreds or thousands of light-years from us. Beyond the Galaxy there is a vast gulf, and then we come to the separate external systems, lying at immense distances. The most famous of them is the Great Spiral in Andromeda, which can be seen with the naked eye as a faint misty patch, and which proves to be a galaxy in its own right, even larger than our own. Herschel had suspected something of the sort, and the work of Rosse and Huggins supported his view, though the question was not finally settled until 1923.

Even the Rosse 72-inch did not retain its lead for long. Each decade saw the arrival of newer and larger telescopes. In

1917 came the 100-inch reflector at Mount Wilson, which remained the greatest in the world until 1948, when it was surpassed by the 200-inch at Palomar.

Whether the Palomar reflector will remain the largest optical telescope to be erected on the Earth's surface remains to be seen. The Russians have announced that they are building a 236-inch, but there is little definite news of this as yet. However, the most pressing need is for more telescopes of the 100–200 inch aperture range, since at present those that do exist are grossly overworked, and cannot possibly carry out all the research that is needed. Incidentally, it should be noted that Britain's famous Greenwich Observatory, now at Herstmonceux, has now a large telescope of its own, the "Isaac Newton" 98-inch.

During the past few decades there have been remarkable developments in other branches of astronomy. The first dates from the early 1930's, when an engineer named Karl Jansky, working for the Bell Telephone Company, was investigating problems of "static" and found that he was picking up radio waves from the sky. This was the beginning of radio astronomy, which has now come so very much to the fore.

Radio telescopes are not in the least like optical telescopes, and they do not produce visible pictures of the objects under study; one cannot look through them, as some earnest inquirers fondly believe! They are designed to collect the long-wavelength radiations coming from space, and they are of many different designs. The most famous radio telescope is probably the 250-foot steerable "dish" at Jodrell Bank, in England, but each design is tailored to suit its own special needs. I am not a radio astronomer, but electronically-minded amateurs will certainly find plenty of scope. Grote Reber, who built a "dish" before the war and was probably the first true radio astronomer, was an amateur.

Associated with radio astronomy is radar astronomy, which involves the transmission of pulses of energy, which are "bounced back" off remote bodies; the echo is picked up, and valuable information gained. It is in this way, by bouncing radar pulses off the planet Venus, that we have obtained the best value for the astronomical unit, or Earth-Sun distance. But radar astronomy is not an amateur pursuit, and I do not propose to follow it up here.

Finally, we come—of course—to rocket astronomy, which had its origin in amateur work before the war, but which became scientifically important only in the years following 1945. To give a full account of all that has happened would take many pages, but it is enough to say that rockets were first used to explore the upper atmosphere, after which came the age of earth satellites—the first, Russia's Sputnik I, went up on October 4, 1957—and lunar and planetary probes. Vehicles have been sent round the Moon, to the Moon, and past Venus and Mars. They have provided information that could not possibily have been gained in any other way. To take just one case: it was in July 1965 that Mariner IV by-passed Mars, and sent back television pictures showing that Mars, like the Moon, is a cratered world. These Martian craters are quite beyond the range of an Earth-based telescope.

I am writing these words in September 1967. By the time that they appear in print, much may have happened; there may be a fully-fledged astronomical observatory circling the Earth, there may be a rocket transmitting data from the region of Mars, there may even be a man on the Moon—though I am pessimistic enough to believe that the lunar journey will not be practicable for some time yet. But in spite of everything, there is still work which the amateur can do. He is more restricted than he used to be, but at least the fascination of astronomy remains as strong as ever.

Chapter Three

TELESCOPES AND OBSERVATORIES

THE CASUAL SKY-WATCHER will be able to give himself hours of enjoyment with the help of nothing more than a pair of binoculars. If he takes the trouble to learn the patterns of the constellations, he will be able to find dozens of double stars, coloured stars, clusters and nebulæ, and he will have no difficulty in tracing the slow movements of the planets against their starry background. Some branches of work, such as the recording of Polar Lights, can be done without any equipment at all.

On the other hand, most of those who feel really drawn towards astronomy will want to obtain some kind of a telescope. No drawing or photograph can give any real idea of the beauty of the lunar mountains, the rings of Saturn, or the myriads of stars in a rich cluster, any more than a rough copy can convey the power and beauty of the Mona Lisa.

A few observatories, such as those at Preston, have "open nights", when members of the general public are allowed to go and look through a powerful telescope. This is admirable, but there is always a queue, and the best that can be done is to have a quick glimpse at some famous object such as a star-cluster or a planet. Large instruments are always busy upon definite programmes, and normally they cannot be made available to amateurs; it would be unreasonable to expect anything of the sort, particularly as a hurried observation is worse than useless. Therefore, the beginner who wants to undertake telescopic work has to obtain equipment of his own. So far as choice is concerned, everything depends upon the interests and the financial resources of the observer. Let us make clear, at the outset, that proper astronomical telescopes are not cheap. Moreoever, very small telescopes are of little value for real work—some that I have seen are less effective than good binoculars.

The beginner has the choice of depending upon binoculars,

or making or buying a telescope. The equipment to be selected must depend upon the interests and the financial resources of the user.

The refracting telescope, basically similar to the tiny instrument made by Galileo, is the usual form. The rays of light coming from the object under observation are collected by a lens or object-glass, which bunches the rays together and brings them to focus. The image produced is then enlarged by another lens, known as the eyepiece. All the actual magnification is done by the eyepiece, and various eyepieces can be fitted to the same telescope.

This seems simple enough, but there are complications. For instance, the eyepiece is generally not a single lens, but a group of lenses held in a casing. The final view will be upside-down, unless deliberately corrected, but this does not matter in the least; in all astronomical photographs and drawings the south is at the top of the picture, with west to the left.

Even the object-glass is not a single lens, and the reason for this is rather interesting. As Newton discovered, what we call "white" light is made up of all the colours of the rainbow, from red to violet. Light may be considered as a wave motion, and the distance from one crest to the next is called the wave-length (Fig. 5). Red light has a longer wave-length than blue or violet, and the result is that the

Fig. 5. Wave-length.

object-glass does not bend it so much. The difference in the amount of bending or "refraction" means that the red rays are brought to focus at a greater distance from the object-glass (Fig. 6). This causes trouble, and the image of a bright object will appear to be surrounded by false colour.

Newton failed to find the remedy, and it was for this reason that he abandoned refractors altogether. Actually, there is at least a partial answer. Modern object-glasses are made up of several lenses, composed of different kinds of glass whose chromatic properties tend to lessen the trouble. The effect can never be eliminated, but it can be very much reduced.

A refractor is classified by the diameter of its object-glass. A "3-inch" has an object-glass 3 inches across, and so on. The largest refractor in the world, that of the Yerkes Observatory, is a 40-inch.

The distance between a lens and its focal point is known as its "focal length", and this length divided by the diameter of the object-glass gives the "focal ratio" (usually abbreviated to "f/ratio"). For instance, I have a 3-inch refractor with a focal length of 36 inches. The f/ratio is therefore $36 \div 3$, or 12. The eyepiece combination has its own focal length, and the magnification obtained depends on the ratio of the focal length of the

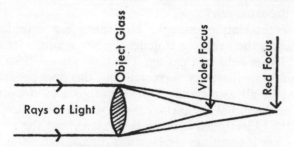

Fig. 6. Unequal refraction. The difference between the refraction of red and violet light has been very much exaggerated, for the sake of clarity.

eyepiece to that of the object-glass. In the case of my own f/12 refractor, an eyepiece of focal length $\frac{1}{2}$ inch will give a magnification of $36 \div \frac{1}{2}$, or 72 diameters—usually written, for short, as "$\times 72$". With an object-glass of focal length 48 inches, the same eyepiece would give a power of $48 \div \frac{1}{2}$, or 96.

It might therefore be thought that the way to get the best out of an eyepiece would be to use it with an object-glass of long focal length. Unfortunately this introduces other troubles, and the only solution is to strike a happy mean.

Naturally, a large object-glass will collect more light than a smaller one. Suppose that I use a very short-focus eyepiece, say $\frac{1}{20}$ inch, upon my 3-inch refractor? The magnification will be $36 \div \frac{1}{20}$, or 720. Yet the image will be so faint that nothing will be made out. The small object-glass, only 3 inches across, is quite unable to collect enough light to satisfy so powerful an eyepiece. If I want to use a magnification of 720, I must buy a larger telescope.

Lens-making can be carried out only by a professional worker, and if the amateur wants to possess a refracting telescope of any size he has no alternative but to buy it. This is not the case with the reflector, and anyone with patience and a certain amount of manual skill can make himself a very adequate instrument.

Newton's arrangement is shown in Fig. 7. Here the light from the distant object passes straight down an open tube until it strikes a mirror at the bottom. This mirror is shaped so as to reflect the rays back up the tube, directing them on to a smaller mirror called a flat. The flat is placed at an angle,

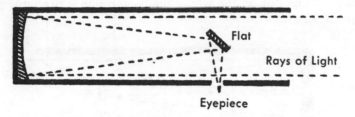

Fig. 7. Principle of the Newtonian reflector. For the sake of clarity, the curve of the main mirror has been much exaggerated.

and sends the rays to the side of the tube, where they are brought to focus and are magnified by an eyepiece in the ordinary way. With a Newtonian reflector, therefore, the observer looks into the side of the tube instead of up it. Of course, the flat prevents some of the light-rays from reaching the main mirror at all, but the loss is not serious, and in any case there is no way of avoiding it.

There is one great advantage in getting rid of the object-glass. A mirror reflects all colours equally, and so the troublesome colour fringes do not appear. For this reason, colour estimates with a reflector are a good deal more reliable than those made with the help of a refractor.

A reflector is classified according to the diameter of its main mirror. However, we must be careful when comparing mirrors with lenses; inch for inch, the lens will give a better result. A 6-inch refractor is appreciably more effective than a 6-inch reflector, so that it can be used with an eyepiece of higher magnification.

Generally speaking, small and moderate reflectors have focal ratios of from f/7 to f/9. There are good reasons for this, but to enter into a full discussion would be beyond our present scope. Nor need we do more than mention the other types of reflecting telescopes; the Gregorian and the Cassegrain, in which the light is reflected back through a hole in the main mirror (Fig. 8), and the Herschelian, in which the main mirror is tilted so as to dispense with the flat altogether (Fig. 9). Gregorians and Herschelians have marked disadvantages, and the amateur will be wise to avoid them. The Cassegrain has many virtues, but on the whole the Newtonian is probably the best for amateur use.

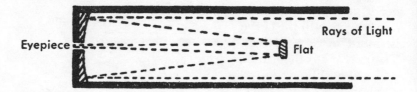

Fig. 8. Principle of the Cassegrain reflector. The "flat" is convex, and is just in front of the point of focus of the main mirror. In the Gregorian reflector, the "flat" is concave, and is placed just beyond the point of focus of the main mirror. The Gregorian gives an erect image.

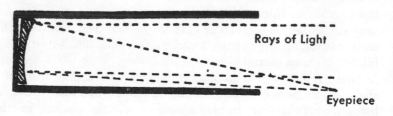

Fig. 9. Principle of the Herschelian reflector. This form of reflector is now virtually obsolete.

Obviously, the performance of a reflector depends entirely upon its main mirror. The surface is coated with a layer of silver, aluminium or rhodium to make it highly reflective, but this is only part of the story. The shape of the curve must be extremely accurate, or the images produced will be distorted. The main mirror is consequently the most expensive part of the whole instrument, if it is to be bought in a finished form.

The principle of grinding a mirror into the correct optical curve is to use two disks of glass, at least an inch thick, one of which will turn into the final mirror while the other is merely a "tool". The tool is fastened to a bench, and the mirror placed on top of it, with water and carborundum powder between the two. The mirror is then slid to and fro, while the operator rotates it and also walks round the bench. Clearly, the tool will be worn away round the edge and will thus become convex, while the mirror will be worn away in the middle and will thus become concave (Fig. 10).

This process is easy, and needs merely a good deal of patience until the curve is more or less correct. The mirror has then to be polished and figured, and a moment's carelessness will ruin hours of work. Numerous tests have to be made, and the real difficulty lies in the "figuring", which means producing the correct curve. But it can be done, and making a 6- or 8-inch mirror is within the capabilities of most people. I

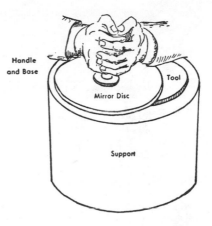

Fig. 10. Grinding a mirror.

know of a fifteen-year-old enthusiast who has made himself a really good 6-inch reflector, and has also built the stand. A list of books giving full instructions will be found in the Appendix on page 318.

The cost of the mirror and tool disks need not be more than $25, and the flat and eyepieces will swallow another $25, while the material for the tube and stand can be bought for a few dollars. In fact, $60 should cover the whole cost. The tube need not even be of rolled metal; it can be a skeleton of lattice construction with a square section, and the only real requirement is that it should be firm.

On the other hand, nobody should set out to grind a mirror without being prepared for a series of setbacks. Difficulties and problems arise at every turn, and there will be moments when the luckless operator feels inclined to hurl his mirror on to the ground and stamp on it. Patience is absolutely necessary—as is the case with almost everything in life.

The construction of a mount is purely a mechanical task. One form, the altazimuth, is shown in Fig. 11. Here the instrument—in this case a 6-inch reflector—is resting in a cradle (A), and is kept in position solely by its own weight. The cradle can be rotated (B), and the telescope can be swung up or down by sliding the rod (C). The top of the rod is fitted with a worm (D), so that by moving the wheel the telescope can be moved very slightly up or down, while the handle (E), attached to a special form of joint, gives a similar slight rotation of the whole telescope. D and E are known as "slow motions". They are not essential, but they are certainly helpful.

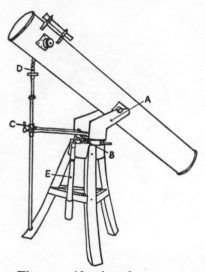

Fig. 11. Altazimuth mount, for a small reflector.

Fig. 12 shows a much simpler mount, this time for a 3-inch refractor. It is simply a tripod, so that the telescope can be moved in any direction; slow motions are not fitted, because they are not necessary for so small an instrument.

The next drawing, Fig. 13, is included as an Awful Warning. It is that appalling contrivance known as the Pillar and Claw Stand, beloved of dealers and despised by serious amateurs. It looks nice, and it is cheap, but it is about as steady as a blancmange. The slightest puff of wind will cause the whole telescope to quiver, and the object under observation will appear to dance about like dice in a shaker. Anyone who buys a small refractor will be almost certain to find that it is mounted upon a pillar and claw. If any real work is to be done, the only solution is to buy a rigid tripod and consign the original stand to the dustbin.

Lastly, we come to the Equatorial Stand (Fig. 14), which is far better than any of those previously described. For a telescope of any size, an equatorial mounting is highly desirable, because the Earth is in rotation.

The spinning of the Earth from west to east means that all the celestial bodies appear to move from east to west. This movement is slow, judged by everyday standards, but when we use a telescope to magnify the size of an object in the sky we also magnify the apparent motion. If the telescope remains stationary, a star or planet will seem to shift steadily across the field until it disappears from view. The telescope has then to be moved until the object is found again. Moreover, there are two motions to be made: up or down ("declination"), and east to west ("right ascension"). Slow motions of the type shown in Fig. 11 provide one answer, and are helpful, but it is irritating to have to fiddle continuously with both wheel and handle. To work in comfort under such conditions, one would need four or five hands.

In the equatorial stand, the "polar axis" is pointed towards the celestial pole, so that only the east-to-west pushing

Fig. 12. Simple tripod mount for a small refractor.

is necessary—the telescope will take care of the up-or-down motion of its own accord. If possible, a driving motor should be attached, regulated so that the telescope moves slowly round at a speed which compensates for the apparent shift of the celestial bodies across the sky.

All these stands can be made. Even the driving clock presents no insuperable difficulties, and in the case of a small telescope a drive can be adapted from an old gramophone motor. The books listed in Appendix XXX will be found to give all the instructions needed.

However, there are some people who are hopelessly clumsy with their hands, or who have no wish to spend hours in the messy, delicate process of mirror-grinding or building a stand. There is no disgrace in this (at least, I hope not!); one cannot do everything, and the solution is to buy a telescope ready made, so that it can be put to use at once.

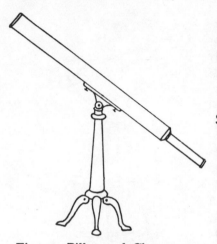

Fig. 13. Pillar and Claw mount for a small refractor. I have nick-named it the "Blancmange" mount, for reasons which should be obvious to anyone who has used it.

What often happens is that the would-be buyer visits a dealer and examines an array of sleek, impressive-looking refractors. He learns that a 3-inch costs about $200, while a 6-inch runs into at least $2000. He is discouraged, and unless he has the sense to ask for advice his astronomical career may end there and then.

Of course, the casual star-gazer who is prepared to spend a substantial sum will gain much pleasure from a 3-inch, even if it is mounted upon a pillar and claw. The instrument will look impos-ing if it is stood in one corner of the library, and it will serve to give adequate small-scale pictures of the Moon, the satellite system of Jupiter and the rings of Saturn, as well as rich star-fields in the Milky Way. However, few people want to spend $200 or more for the sake of occasional amusement, and a smaller instrument, such as a 2-inch refractor, is of little use astronomically. Moreover, even a 2-inch costs upwards of $100 if bought new, plus extra sums for essentials such as stands, focusing arrangements and eyepieces. A word of warning is necessary here. It is of no use whatsoever trying to use any telescope for astronomical purposes unless it is fitted with a stand of some kind. Any sort of stand is better than nothing.*

Though new refractors are expensive, it is sometimes possible to pick up a cheap second-hand 3-inch. Anyone who is prepared to make regular visits to junk shops stands a good chance of finding such an instrument eventually, and it is also worth while to keep a close watch upon the advertisement columns of newspapers and periodicals. There is no guarantee of rapid success, but a 3-inch refractor is an excellent instrument for the beginner, provided that it is firmly mounted. Once the user has gained experience, he will be ready to change to something larger.

Refractors are easy to handle, but they are not light, and a 4-inch is the limiting size for portability. A larger instrument needs a permanent home, preferably some kind of run-off shed or observatory. Few 4-inch refractors are to be found second-hand, and in the ordinary way the cost of a new instrument is prohibitive.

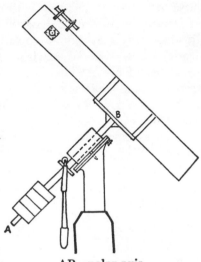

AB = polar axis.

Fig. 14. Equatorial mount.

Reflectors are cheaper, and are much more portable, particularly when fitted with skeleton tubes. Here again the cost of a complete new instrument is rather high, but second-hand reflectors of from 6- to 8-inch aperture can be found quite frequently. The beginner with $150 to spend may indeed

* Some time ago, I had a letter from a beginner who had a 2-inch refractor and was disappointed with its performance. As soon as he mounted it upon an improvised stand, he found that it worked very well.

have the choice between a new 2-inch refractor, or a second-hand 6-inch reflector; obviously, he will do far better to buy the reflector, even if it needs repairing. In my view there is no point in spending much money on a small refractor of aperture 2 or 3 inches, unless it is to be used only for occasional star-gazing.

It is wise to be careful when buying a second-hand telescope, particularly a reflector. It may look perfectly sound, with polished fittings and a beautifully-painted tube; but if the mirror is poor, the performance also will be poor, and defects in a mirror do not always show themselves at first sight. Of course, one way of deciding is to make a practical test upon a star image; but if the telescope lacks a usable stand, or needs adjusting, this may not be possible, in which case the only safeguard is to seek advice from somebody who has a sound knowledge of optics. The beginner who spends pounds upon a second-hand reflector only to find that the mirror is of no use is unlikely to receive much sympathy—nor will he deserve it.

Let us assume, then, that we have managed to acquire a telescope. What care must be taken of it, and what extra equipment shall we need?

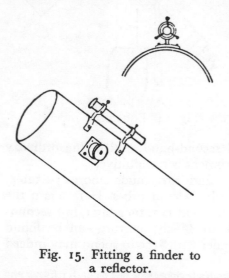

One addition is simplicity itself. A small sighting telescope or "finder" can be fitted, and will be found most useful (Fig. 15). Even a toy telescope will do, and can be attached by Meccano. The advantage of a finder is that it has a large field of view, and will save much time when a faint object is being searched for. The object is simply brought to the centre of the finder field; if the adjustments are correct,

Fig. 15. Fitting a finder to a reflector.

the object will then be visible in the field of the main telescope.

A finder is not strictly necessary; but it is so cheap, and so easy to fit, that it seems a pity not to have one.

More important is the dew-cap, which is simply a short tube which fits over the object-glass end of the refractor in order to prevent dust, dirt and dew from settling on the lens. It can be made from a cocoa-tin lined with blotting paper, or something of the kind, and a cap should always be kept over the object-glass when the telescope is not actually in use (Fig. 16).

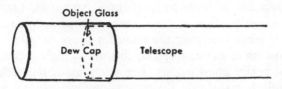

Fig. 16. Dew-cap for a refractor.

If the object-glass needs cleaning, it should be brushed very gently with a camel's-hair brush and then wiped even more gently with a piece of very fine, clean silk or wash-leather. To take the various components of an object-glass apart is most unwise unless the owner has a really good idea of what he is about. All things considered, a small refractor should need little or no attention for years on end, provided that it is not roughly handled. When some major adjustment does become necessary, it will be worth while to take the whole instrument to an expert. It is better to spend a little money on maintenance than a great deal of money on buying a new telescope.

Reflectors need more attention. The main mirror and the flat need periodical re-silvering, and although this can be done at home it does need a good deal of care. It is probably better to have the mirror aluminized, which will give a much longer period before anything further need be done; rhodium coating can also be used. Both mirror and flat should be kept covered with a protecting cap except when actually in use, and yet another word of warning may be timely here. Before using the telescope, uncap the flat before you expose the main mirror. I know of one luckless observer who uncovered the main mirror first—and then dropped the flat cover on to it. He spent the next few months grinding himself a new mirror.

Eyepieces are vitally important, since using a good telescope with a bad eyepiece is like using a good record-player with a bad needle. Theoretically (though not always in practice) eyepieces are made to a standard thread, so that any eyepiece should fit any telescope; but the magnification obtained depends upon the focal length of the mirror or object-glass, so that an eyepiece which yields ×50 on a 3-inch refractor will not yield ×50 on a 6-inch. Moreover, eyepieces are of various types, adapted for different types of telescopes.

It is advisable to have at least three eyepieces. One should give low magnification, for star-sweeping and general views; the second, moderate magnification for more detailed views of planets and some stellar objects; the third, high magnification for use on really good nights. For my 3-inch, f/12 refractor I have found that suitable magnifications are 36, 72 and 144, while for a 6-inch reflector the corresponding powers might be 50, 120 to 180, and 300 to 360. Individual observers are bound to have their own ideas on the subject.

One thing is however important: Do not try to use too high a power. If the image becomes even slightly blurred, change at once to a lower magnification. It may be impressive to say that an observation was made " ×400" or " ×500", but it will often be found that a smaller, sharper picture will yield far more detail.

Let us sum up what has been said. If a telescope is to be bought, it will be far better to search for a moderate reflector than to spend a large sum of money on a portable refractor. Never buy a telescope until you have had an expert opinion on it, since although it may look sound it is quite likely to be useless. Most important of all, do not trust your own judgement unless you are sure that you are really competent. Search in second-hand shops and advertisement columns until you find something that you think will suit you; see it; have it checked; and if you are satisfied, buy it.

A favourite mistake is to poke a telescope through the bedroom window in the expectation of seeing fine detail on the Moon or a planet. Actually, good results can seldom or never be obtained in any such way. The temperature of the room is almost certain to be higher than that outside, and the resultant air-currents will destroy the sharpness of the image. Moreover,

46

there are other hazards. In my very early days I tried to observe from the warmth of my bedroom, and dropped an eyepiece fifteen feet on to a gravel path.

A portable telescope on a tripod is easy to carry about, but a reflector larger than 8-inch aperture or a refractor larger than 4 inches is generally too heavy to be moved far. This means that it must be set up in one permanent position, and should be protected against the weather.

If the telescope is carefully painted it may not come to much harm, and one easy solution is to provide it with a waterproof cover. A run-off "observatory" is simple to build; it can be run on rails, so that when the telescope is to be used the whole shed can be rolled back out of the way. It may be either centrally divided or else made in one section.

A proper observatory is built in the form of a dome, one section of which can be removed, either by taking it out or by swinging it back on hinges. The dome can be revolved, generally by being pulled round with ropes. Here again the problems are purely practical, but a dome is by far the best form of observatory, partly because it protects the telescope completely and partly because the observer is shielded from the chill breeze of a winter night.

The great observatories of the world are among mankind's finest creations. Pride of place must go to that on Palomar Mountain, in California, where the main instrument is the 200-inch Hale reflector, named in honour of Professor George Ellery Hale, who was the moving spirit behind its construction. So far, the 200-inch remains in a class of its own; its nearest rival is the 120-inch at the Lick Observatory, also in America.

A vast reflector costs a fabulous sum, not only because of the optical parts but also because of the extra equipment needed. An observatory such as Palomar is almost a city in itself. There are instruments of all types, each in its own dome; laboratories; dark rooms; living quarters and lecture theatres. At Palomar, the 200-inch may be the main instrument, but there are many others as well.

Most great telescopes are set up in observatories high above sea-level. This is because the atmosphere, so necessary to life, is a positive handicap from the astronomer's point of view. Not only is it dirty, but it is also turbulent, so that using a high magnification will result in violent unsteadiness of the image. The densest part of the air-mantle is concentrated near the Earth's surface, and by climbing as high as possible we can reduce the disturbance, though we can never really cure it.

Greenwich Observatory, so familiar a name to all of us, has had to contend with extra problems. When Sir Christopher Wren designed the original structure, in the reign of Charles II, Greenwich was a village well outside London; there were few artificial lights, and the air was clear and smoke-free. Nowadays the situation is very different. London's tentacles have stretched out, and Greenwich has become a suburb. Electric lamps cause a glare across the sky, while the smoke from a thousand factory chimneys causes an everlasting pall.

Modern Greenwich is, in fact, no place for a large telescope. When it was proposed to build a 98-inch reflector, only slightly smaller than the Mount Wilson colossus, a decision had to be made. To set up a vast instrument in a smoky atmosphere would be pure folly, and so it was agreed to move the whole observatory to Herstmonceux, near the little Sussex town of Hailsham, where seeing conditions were still relatively good. The war held matters up, and the move took a very long time, but by the beginning of 1957 there was little left at Old Greenwich apart from historical relics. Perhaps the most interesting of these relics is the Octagon Room, where John Flamsteed worked away at his famous star catalogue. This at least is open to the public, and is well worth a visit. Still to be seen there is the "transit telescope" built by Edmond Halley, who succeeded Flamsteed and became the second Astronomer Royal. There is also an interesting "Herschel Room". Meanwhile, the 98-inch reflector has come into use at Herstmonceux.

Making a 200-inch mirror is difficult enough, but making a 200-inch object-glass would be quite out of the question, even if it could be mounted satisfactorily. The largest refractor in the world is the 40-inch at Yerkes, in the United States, while the largest in Europe is the 33-inch at Meudon, between Paris and Versailles. The Meudon instrument itself is over 70 feet long; I have had the privilege of making extensive lunar observations with it, and so have personal experience of its high quality. The 22-inch refractor at the Pic du Midi, in the French Pyrenees, is another great telescope, and since it lies

48

at a height of 10,000 feet it can be used under conditions of great clarity. It is probably true to say that the lunar and planetary photographs taken there are the best that have ever been produced. At the Lowell Observatory, at Flagstaff in Arizona, there is the famous 24-inch refractor, now used largely on the Moon-map project.

A visit to an observatory is not an experience to be missed, and it is a great pity that the chance comes to so few people. Yet we must not look down upon the humble run-off shed covering a modest instrument. Astronomy is open to all.

Chapter Four

THE SOLAR SYSTEM

SOME PEOPLE REFUSE to take an interest in astronomy simply because they are frightened of it. They cannot appreciate distances of millions of miles; they cannot believe that each star is a sun, and their minds remain firmly anchored to the Earth.

This point of view is commoner than might be imagined, and part of the difficulty originates from the vast scale of the universe. Nobody can really picture "a million miles", and the tremendous heat of the Sun's interior is equally beyond the human brain. The best way to give some account of scale is to visualize a model, which will at least put our ideas in some sort of order.

The Solar System in which we live is made up of one star (the Sun), nine major planets, and numerous bodies of lesser importance, such as the moons or "satellites", the minor planets, the meteors and the comets. Returning to the model discussed on page 30, we imagine that the Sun has become a globe only 2 feet in diameter, so that we can put in the rest of the planets on the correct scale. Mercury will become a grain of mustard seed 83 feet from the 2-foot Sun; Venus, a pea at 156 feet; the Earth, another pea at 215 feet; Mars, a pin's head at 328 feet; Jupiter, an orange at $\frac{1}{5}$ of a mile; Saturn, a tangerine at $\frac{2}{5}$ of a mile; Uranus, a plum at $\frac{4}{5}$ of a mile; Neptune, a slightly smaller plum at $1\frac{1}{4}$ miles; and Pluto another pin's head, with a maximum distance of 2 miles. The nearest of the ordinary stars will then lie 10,000 miles off, which gives us a good idea of how isolated the Solar System really is.

There is a great deal of difference between a 2-foot globe and an orange, and so even Jupiter, largest and most massive of the nine planets, is by far inferior to the Sun. The Sun is in fact the absolute ruler of our system; it controls the movements of the planets, and the planets depend entirely upon solar heat and warmth. No planet has any light of its own. Even Venus, the glorious "evening star" which can shine down like a small lamp and can even cast a shadow at times, is in itself a dark, non-luminous body.

One thing is evident from our scale model: the planets can be divided into two well-marked groups. The inner group is made up of four small and comparatively close-in worlds, Mercury, Venus, the Earth and Mars. Then comes a wide gap, followed by the four giants, with that curious little world Pluto on the very fringe of the Sun's kingdom. Actually, the gulf between Mars and Jupiter is not empty. It is occupied by many thousands of tiny bodies, the Minor Planets or asteroids, which would be mere grains of dust on the scale which we have chosen.

The individual motions of the bright planets have been known since very early times, and the very word "planet" means "wandering star". The ancients also noticed that the planets keep strictly to a certain region of the sky, which they named the Zodiac. The reason for this is that the paths or "orbits" of the planets lie almost in the same plane, so that when we draw a plan of the Solar System upon a piece of flat paper, as in Fig. 17, we are not very far wrong. Consequently, the planets can be seen only in certain directions, and this limitation applies also to the Sun and the Moon. The Sun's apparent yearly path among the stars indicates the "ecliptic".

A good way to make this clear is to imagine that we are standing in a wood, looking at low trees around us. There may be trees to all sides, but no trees will appear in the sky or beneath our feet—because the trees lie in roughly one plane, the plane of the Earth's surface.

As we know, the stars were originally looked upon as mere points of reference. The early astronomers grouped them into constellations, and there are twelve constellations in the the Zodiacal band, which stretches right round the heavens. The most famous of these groups is probably Aries, the Ram. It contains no very bright stars, and is not particularly easy to identify, but in the far-off times when the Chaldæan shepherd-astronomers gazed at the skies during their night watches Aries was the constellation in which the ecliptic cut the "celestial equator", the projection of the Earth's equator upon the celestial sphere. Actually, the point of intersection, or

"First Point of Aries", has moved since then, because of the wandering of the polar point, and has now passed into the neighbouring constellation of the Fishes; but we still keep to the old term.

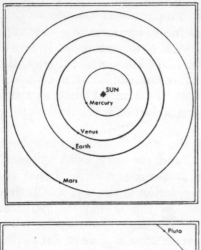

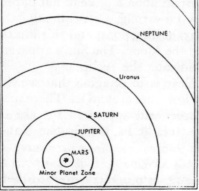

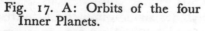

Fig. 17. A: Orbits of the four Inner Planets.

B: Orbits of Mars and the five Outer Planets.

Since the planets are never far from the ecliptic, they are easy to recognize. In any case, Mars (when at its brightest) and Jupiter are so distinctive that they cannot possibly be confused with stars, while Mercury and Venus, which are closer to the Sun than we are, have their own way of behaving. Only Saturn, and Mars when at its faintest, cannot be identified at the most casual glance.

The first astronomer to give a proper description of the way in which the planets move was Johann Kepler. Between 1609 and 1619 he published his three famous Laws of Motion, which are interesting enough to describe in slightly more detail. They are as follows:

Law 1. The planets move in ellipses, with the Sun at one focus.

Law 2. The radius vector (the line joining the centre of the planet to the centre of the Sun) sweeps out equal areas of space in equal times.

Law 3. The square of the sidereal period is proportional to the cube of the planet's mean distance from the Sun.

These may seem rather complex, but really they are quite simple. Law 1 requires no explaining; the only point to bear in mind is that although the orbits of the planets are ellipses, they are of slight eccentricity, and do not depart much from the circular form. It is the other two Laws which sometimes cause beginners to wrinkle their brows.

Law No. 2 is explained by the diagram in Fig. 18. The figure is not to scale, and the orbit of our supposed planet P is much more eccentric than is actually the case with any major planet in the Solar System, but one has to make the diagram

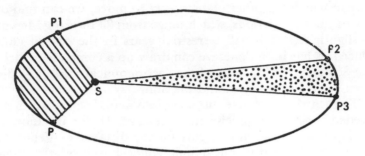

Fig. 18. Kepler's Second Law.

inaccurate in order to make it clear! S is the Sun; P, P1, P2 and P3 stand for the planet in various positions in its orbit round the sun.

Assume that the planet moves from P to P1 in the same time that it takes to go from P2 to P3. Then the shaded area of PSP_1 must be equal in area to the dotted area of P_2SP_3. Since the dotted area is "longer and thinner", it is clear that the planet is moving at its quickest when closest to the Sun.

This fact is vitally important. It can be summed up by the simple rule "The nearer, the faster". The Law does not mean only that a planet moving in an elliptical orbit must travel at a varying speed; it means also that a planet when close to the Sun must move faster than when it is more distant. This is borne out by direct measurement. Mercury, for instance, has an orbit which is definitely eccentric, so that at its closest to the Sun ("perihelion") it is only $28\frac{1}{2}$ million miles away, as compared with $43\frac{1}{2}$ million miles at its farthest point ("aphelion"). The orbital speed varies from $36\frac{1}{2}$ miles per second at perihelion to only 24 at aphelion. The Earth, at the greater distance of 93 million miles, is a comparative sluggard, and has an average rate of a mere $18\frac{1}{2}$ miles per second.

The Third Law leads to some equally important conclusions. The "sidereal period" of a planet, the period taken to complete one revolution round the Sun—the planet's "year"—is linked with the actual distance from the Sun, and if we know the one we can find the other.

The Earth's sidereal period is $365\frac{1}{4}$ days. By studying the way in which the other planets seem to move, we can find out their respective periods, which range from 88 days for Mercury to slightly less than 248 terrestrial years in the case of Pluto. Once this has been done, we can draw up a complete model of the Solar System in terms of the "astronomical unit", the distance between the Earth and the Sun.

To turn these relative distances into actual miles, all that is needed is any one precise measurement. If, for instance, we could obtain an accurate figure for the distance of Venus, the length of the astronomical unit could be calculated. Since 1961 radar methods have been used by both the Americans and the Russians, the general principle being to "bounce" an energy pulse off Venus, time the delay before the "echo" returns, and then calculate the distance travelled—remembering that a radar pulse, like visible light, moves at 186,000 miles per second. It is now thought that the mean Earth-Sun distance amounts to approximately 92,868,000 miles.

The Moon, which revolves round the Earth,* is of special interest to us. Everyone is familiar with its monthly phases, from new to full and back again to new, but not everyone is sure how they are caused. Some people still believe that they are due to the shadow of the Earth, but the true explanation is far simpler.

The Moon is a dark body, shining only by reflected sunlight. As the Sun can light up only one half of the lunar globe at a time, the other half must be non-luminous, and therefore

* Actually, the Earth and Moon revolve round their common centre of gravity; but as this point lies within the terrestrial globe, the plain statement that "the Moon revolves round the Earth" is good enough for most purposes.

invisible. In Fig. 19, the Moon is shown in four positions in its monthly journey—M1 to M4. At M1, the dark side is turned towards us; since this does not shine, the Moon is invisible, or new. As the Moon moves on towards M2, a little of the day hemisphere starts to turn in our direction, and we see the familiar crescent shape; by the time M2 is reached, half the sunlit side is presented, and the Moon is at half phase. (Rather

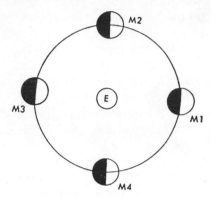

Fig. 19. Phases of the Moon. S—Sun; E—Earth; M1 to M4—the Moon in four different positions in its orbit. Not to scale.

confusingly, this is termed First Quarter—because the Moon has completed roughly one quarter of its orbit from new to new.) Between M2 and M3 the appearance is "gibbous", between half and full, and by the time M3 is reached the Moon shows us the whole of its day hemisphere. After Full, the phase wanes once more, to half-moon at M4 (Last Quarter) and then crescent, until M1 is reached at the next new moon.

Clearly, the Earth's shadow has nothing to do with these phases. It is true that when the Moon is full (M3) and the three bodies are perfectly lined up, the shadow of our globe does fall across the Moon, causing a lunar eclipse; but eclipses do not occur every month, because the Moon's orbit is somewhat tilted with respect to ours.

The lunar phases must have been known since the dawn of history, but it was not until the invention of the telescope that Venus and Mercury were found to behave in a similar way.

The phases of Venus, first detected by Galileo, are explained by Fig. 20. E represents the Earth, which is assumed to be stationary (really, of course, it is moving round the Sun all the time, but this makes no difference to the illustration); S the Sun, and V1 to V4 Venus in four different positions. Since Venus is closer to the Sun than we are, and moves more quickly, it completes one circuit in only 224·7 terrestrial days.

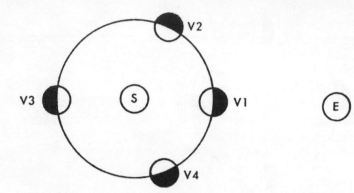

Fig. 20. Phases of Venus. S—Sun; E—Earth; V1 to V4— Venus in four different positions in its orbit. Not to scale.

At V1 the Earth, Venus and the Sun are in a straight line, with Venus in the middle. The night side is then turned towards us, and Venus is new, so that it cannot be seen at all. This position is known as "inferior conjunction". Occasionally the alignment is perfect, and Venus can be seen as a black spot against the solar disk; but since Venus too has a tilted orbit, these "transits" are rare. The next will not occur until the year 2004.

As Venus moves on towards V2, we start to see the sunlit side. The planet appears in the morning sky as a slender crescent, becoming brighter and brighter as it draws away from the line of sight with the Sun. At V2 the three bodies form a right-angled triangle, so that Venus appears as a half disk. It then rises some hours before the Sun, and is a splendid object in the east before dawn. The technical term for this is "Western" or Morning Elongation.

As it travels towards V3, Venus changes from a half into a

gibbous disk, and draws back towards the direction of the Sun so that it grows steadily less conspicuous. By the time it has reached V_3, it has ceased to be visible except during the hours of daylight. It is then at "superior conjunction", and since it lies almost behind the Sun it is not easy to find even with a telescope.

After passing superior conjunction, Venus makes its appearance low down in the evening sky, shrinking gradually to a half as its angular distance from the Sun grows. It reaches eastern elongation at V_4, and is then at half-phase once more, after which it narrows to a crescent as it returns to inferior conjunction at V_1.

The "synodic period" of Venus, the interval between one inferior conjunction and the next, is 584 days, though this may vary by as much as four days either way. The interval between its appearance at V_4 and that at V_2 is about 144 days, while 440 days are needed for the much longer interval between the appearance at V_2 and that at V_4.

Venus is of course at its closest to the Earth at inferior conjunction. The distance is then reduced to about 24 million miles, about a hundred times as great as that of the Moon; but as the dark side is then almost wholly presented, we cannot see the planet at all. When the disk is almost fully illuminated, Venus is a long way away. It is in fact a most infuriating object to observe.

Mercury behaves in the same manner as Venus; but since it is smaller, as well as being closer to the Sun, it is much less easy to study. It is never conspicuous to the naked eye, and only at favourable elongations can it be seen glittering near the horizon like a star. This is interesting, in view of the fact that many people believe that planets cannot twinkle. It is true that a planet, which shows a definite disk, twinkles much less than a star, which appears only as a minute point of light; but when a planet is low down, and thus shining through a dense layer of atmosphere, it may twinkle violently. This is particularly so in the case of Mercury.

The remaining planets lie beyond the Earth in the Solar System, and cannot appear as halves or crescents. Mars is shown in the diagram (Fig. 21), and is typical of all the rest.

Let us start with the Earth at E_1 and Mars at M_1. The

57

Sun, the Earth and Mars are lined up, with the Earth in the middle; Mars is therefore directly opposite the Sun, and is at "opposition". One year later, the Earth will have completed one revolution, and will have arrived back at E1; but Mars, moving more slowly in a larger orbit, will not have had time to get back to M1. It will have travelled only as far as M2, and will lie on the far side of the Sun, badly placed for observation. The Earth has to catch it up, with Mars moving onwards all the time, and on an average 780 days elapse before the three bodies are lined up again. The 780-day interval between successive oppositions is therefore the synodic period of Mars.

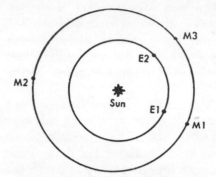

Fig. 21. Oppositions of Mars. E1 and M1—positions of Earth and Mars at the oppositions of September 1956; E2 and M3—positions at the opposition of November 1958. There was no opposition of Mars in 1957 (see text).

The giant planets are much more remote, and move so much more slowly that the Earth takes less time to catch them up. Jupiter's synodic period is 399 days, while in the case of far-off, sluggish Pluto the period is 366¾ days. After having completed one circuit of the Sun, the Earth has to travel on for only an extra day and a half before it catches up with Pluto.

Each of the nine major planets has its own characteristics. The members of the inner group are small, solid bodies; all have atmospheres, though that of Mercury is negligible; and there is a strong probability that low forms of plant life flourish on Mars. The giants, however, are built up a very different pattern. When we look at Jupiter or Saturn, what we see is not a solid, rocky globe, but the outer layer of a deep "atmosphere" made up of poisonous gases. Pluto presents problems of its own, but since it is so faint and so far away it is not of much interest to the amateur.

Most of the planets have moons, or satellites. The Earth, of course, has one; Jupiter boasts of twelve, Saturn ten, Uranus five, and Neptune and Mars two each, while Mercury, Venus and Pluto do not seem to possess any. A small telescope will show the brightest of these satellites, and a few exceptionally keen-sighted persons are said to have seen one or two of the chief satellites of Jupiter without a telescope at all.

The minor planets, or asteroids, swarm in the wide gap between the orbits of Mars and Jupiter. All are dwarf worldlets, less than 500 miles in diameter, and only one (Vesta) is ever visible without optical aid. Even with a telescope, they look remarkably like small stars, and the only way to identify them is to watch them from night to night, until their slow movement across the starry background betrays their true nature.

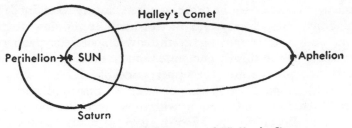

Fig. 22. Orbits of Saturn and Halley's Comet.

The remaining members of the Sun's family are much less substantial. Particularly interesting are the comets, which have been termed the stray members of the Solar System. Most of them move in elliptical orbits, but their orbits are much more eccentric than those of the planets. Fig. 22 shows the path of a periodical comet (Halley's) as compared with the orbit of Saturn. Nor is a comet a solid body; it is made up of a swarm of particles contained in an envelope of very thin gas. A famous astronomer once called comets "airy nothings", and though they are not "airy" in the usual sense of the word they are certainly flimsy. Comets may be of immense size, but they are of negligible mass, and they are of course completely harmless, even though they still strike terror into the hearts of some of the Earth's backward races.

The ghostlike nature of a comet means that it can be seen only when it is fairly close to the Earth and to the Sun. Halley's Comet—named after Edmond Halley, Flamsteed's successor at Greenwich, who was the first to discover that it revolves round the Sun—has a period of 76 years, but for most of that time it is too faint to be seen. It last came to perihelion in 1910, in the reign of Edward VII, and will not reappear until 1986. We know where it is, but at present we cannot observe it.

Other comets have shorter periods, and can thus be seen every four or five years. Others, however, have periods of many centuries. Each year brings forth its quota of new faint comets, though not many of them become bright enough to be seen without the aid of a telescope.

Meteors, or shooting-stars, are also members of the Solar System. The name is misleading, since they are not stars at all. They are small pieces of matter travelling round the Sun in elliptical orbits, and in the ordinary way they are too faint to be seen. Sometimes, however, a meteor may come close to the Earth, and if it is moving at the right speed in the right direction it will naturally encounter the Earth's mantle of air. It will then plunge into the upper atmosphere, and will rub against the air-particles, setting up friction; first it will become warm, then hot, and then it will burst into flame, usually burning itself completely away in a matter of seconds and finishing its earthward journey in the form of fine dust.

It is easy to prove that air sets up resistance. If you cup your hand and swing it abruptly, you can feel the pressure; a stick hisses through the air if swished, and the air in a bicycle pump becomes hot when a tyre is being blown up. Small wonder that a meteor, travelling at a tremendous speed, will become violently heated. Above a height of 120 miles or so, the air is of course too thin to cause appreciable resistance.

Such is the Solar System. It contains bodies of all kinds, from the vast, intensely luminous Sun down to tiny particles of interplanetary dust, and even though it may be unimportant in the universe as a whole it is of supreme importance to ourselves.

Yet there has been a decided change of attitude during the past few years. Previously, professional astronomers were busy with their studies of the greater universe, and paid scant attention to the surfaces of the Moon and planets; so far as the lunar craters or the Martian deserts were concerned, the amateur was left a clear field, which is why so much of our present-day knowledge is based upon amateur work. Then, in 1957, the first earth satellite was launched, and the Space Age began. From being a remote body of interest only to a few enthusiasts, the Moon became officially accessible, with Mars and Venus next on the astronautical list. One must admit, with regret, that military planning had something to do with the change of view. A rocket that can launch a lunar probe can also launch a nuclear bomb.

At any rate, bodies such as the United States Air Force began to show a lively interest in the Moon, and official photographic programmes were started. They are still in progress, and are being conducted not only with great skill, but with the best possible equipment and with ample funds. This also applies, though with less force, to Mars. The amateur, necessarily working with modest telescopes, is no longer alone in his work.

This is, of course, a good thing, because it leads to a better understanding of the Moon and planets. The only point to be borne in mind is that the 1967 amateur must be much more selective in his observational programme than the 1957 amateur had to be. There is little to be gained nowadays from drawing areas of the Moon with a modest telescope—unless there is some special reason for so doing; and a sketch of Mars with, say a 6-inch reflector appears somewhat limited when compared with a photograph from Mariner IV. All the same, amateur observations of the bodies in the Solar System are still of value if properly conducted. Moreover, rockets are still limited in their own range. The Moon, Mars and Venus may well be reached within the next century or so—the Moon much earlier than that—but it will be a very long time indeed before the first explorers are able to enjoy close-up views of the belts and moons of Jupiter or the superb ring-system of Saturn.

Chapter Five

THE SUN

STUDYING THE SUN calls for methods different from those used in any other branch of astronomy. In other cases, the main problem is to collect as much light as possible; with solar observation there is plenty of light available, but it is highly dangerous to look directly at the Sun's disk using a telescope, as the eye of the observer is certain to be damaged.

The Sun's diameter is 865,000 miles, 109 times that of the Earth. But though the solar globe could contain over a million bodies the size of our own world, it does not contain the mass of a million Earths. Only 332,000 Earths would be required to make one body with the mass of the Sun. This means that the Sun is less massive than one might expect from its size, and that the mean density is less than that of the Earth —in fact, only 1·4 times as great as that of water.

Of course, this is not the uniform density throughout the solar globe. Density increases with depth. Near the centre of the Sun, the material is denser than steel, even though it is still technically a gas, whereas the outermost parts of the Sun are more rarefied than the best vacuum we can produce in terrestrial laboratories.

The gravitational force that would be felt by a man standing on the surface of a globe depends upon two factors, the mass and the size. Taking the Earth's surface gravity as unity, the surface gravity of another body can be found by dividing the mass by the square of the radius. For the Sun, these figures are respectively 332,000 and 109, so that the surface gravity is 332,000 divided by 109 squared, or 28. A man who weighs 14 stone on Earth would weigh 2½ tons if he could be taken to the surface of the Sun, so that he would not even be able to stand upright; he would be crushed by his own weight. However, there is no prospect of life, human or otherwise, surviving on the solar surface. Even one of the tough, grizzled space-captains of science fiction would feel uncomfortably warm there, since the temperature amounts to 6,000 degrees.

The great size and the low density mean that the Sun cannot be a solid body like the Earth. It is in fact made up entirely of gas, though deep down inside the globe this gas is under tremendous pressure—at least a thousand million atmospheres—and behaves therefore in a decidedly un-gaslike manner judged by our normal standards.

Telescopic views of the Sun do not tell us much more. Interesting features can be seen, such as the dark spots and the brighter patches or faculæ, but for more serious studies it is necessary to use special instruments. Some of these can be made by the skilled amateur, but to describe them would be beyond the scope of the present chapter, and all that can be done here is to summarize the results obtained by them.

Newton was the first to explain the breaking-down of white light into its constituent colours. What he did was to cut a small hole in the shutter of his window, so that only a narrow beam of sunlight could pass through. This beam entered a glass prism, and the resulting rainbow or spectrum was spread out on the far wall. Later, Newton improved the experiment by using a slit instead of a hole, and by putting a lens between the prism and the wall so that he could bring the colours to a sharp focus.

Newton never took his investigations much further, probably because his prisms were of poor quality glass. The next development was due to Fraunhofer, who returned to the problem in 1814, and who found that the spectrum of the Sun is crossed by numbers of dark lines of different degrees of intensity. It is now known that each of the Fraunhofer Lines is due to the effect of one definite substance, and this is the basis of all solar and stellar spectroscopic work. One substance (such as iron) may produce many characteristic lines.

It may be added that dark lines had been seen in 1802 by a British scientist, Wollaston. Wollaston did not however realize their importance, and thought that they merely marked the boundaries of the different spectrum colours, so that the main credit must go to Fraunhofer.

All matter in the universe, whether in the Earth, the Sun or the remotest star, is made up of different combinations of a small number of fundamental "elements". There are 92 elements in all, hydrogen being the lightest and uranium the heaviest, and since they form a complete series there is no chance of our having missed one. No new elements can exist, because there is no room for them in the sequence; one might as well try to fit an extra integer between 7 and 8, or a new musical note between F-sharp and G. (It is true that various extra elements have been made artificially in recent years, but these "lead on" from the end of the sequence, and probably do not occur naturally.) We can thus be certain that each Fraunhofer Line is due to an element or group of elements already known to us.

When observed with the aid of a prism or spectroscope, the bright surface of the Sun, the photosphere, gives the bright rainbow studied by Newton. Above this is a layer of incandescent vapour, extending upwards for perhaps 10,000 miles. On its own, this vapour would give not a rainbow, but a number of bright isolated spectrum lines. However, there is the bright background to be taken into account, and the result is that instead of appearing bright, the lines emitted by the upper vapour are "reversed", and seem to be dark. This "reversing layer" is the outer envelope, or chromosphere.*

The dark lines give us a key to the elements responsible for them. The spectra of the various elements have been studied in terrestrial laboratories, and the positions of the lines are known with high accuracy, so that all we have to do is to compare the laboratory lines with those visible in the solar spectrum. If a solar line corresponds to a laboratory line of sodium, we can prove that there is sodium in the Sun. In this way nearly 70 of the 92 natural elements have already been identified.

We are now in a position to examine the structure of the Sun itself. Near the centre of the globe, the pressure is tremendous, while the temperature is terrifyingly high—something like 14 million degrees Centigrade,† which is beyond our comprehension. It is here that the production of energy is going on,

---

* Many books differentiate between the "reversing layer" and the "chromosphere", but there is no justification for this. The whole chromosphere is a reversing layer, though all the solar elements occur in the lower part of it, so that this part gives the most complete bright-line spectrum.

† In general, I have given stellar temperatures in degrees Centigrade and planetary temperatures in Fahrenheit. Some may object to this practice, but it is very easy to convert one scale into the other. It involves nothing more frightening than simple multiplication and division.

and the inner region has aptly been termed "the solar power-house".

The visible surface of the Sun, the photosphere, is the region where the solar gases become thin enough to be transparent. The bright rainbow spectrum originates in the photosphere, and here too we meet the curious dark patches which are known, rather misleadingly, as sunspots.

Above the photosphere we come to the chromosphere, or "colour sphere", made up largely of hydrogen gas. Except during a total solar eclipse, it cannot be seen except by using special instruments, since the intense glare from the photosphere hides it completely. Finally, beyond the chromosphere, we come to the extended outer atmosphere of the Sun, known as the corona.

The most interesting objects in the chromosphere and corona are the prominences. These are masses of glowing vapour, composed mainly of hydrogen, helium and calcium. Some of them have been known to climb to a million miles above the bright surface, and they move so rapidly that there must be occasions upon which parts of them escape from the Sun altogether, leaking away into space.

The difficulty of observing the prominences, the corona and other interesting high-level features is that the equipment needed is expensive to buy, and not too easy to make. The chief instruments of the serious research worker are the spectro-heliograph, the spectrohelioscope and the monochromatic filter, all of which enable the observer to study the Sun in the light of one particular element only—usually hydrogen or calcium. The prominences and the chromosphere may be studied at any time with such instruments, together with other features such as the curious dark patches known as flocculi.

Sunspots, which are almost equally interesting, can however be seen with any small telescope; some of them become large enough to be visible with the naked eye, and records of them go back to Ancient China. It is a fascinating pursuit to track them as they drift slowly across the Sun's disk, and to watch their shapes change from day to day.

At the moment we have to confess that although we know a great deal about the way in which sunspots behave, we are still uncertain as to their origin and significance. A spot may be

65

described as a relatively cool patch on the photosphere, so that it emits less light than the surrounding surface. "Cool", however, is here used in the solar and not the terrestrial sense; the mildest part of a spot still has a temperature of some 4,000 degrees, but the difference between this and the normal photosphere is enough to make the spot appear dark. If seen by itself, it would however glow with a brilliance much greater than that of an arc-lamp, so that it would be a grave mistake to describe a sunspot as "black".

A large spot is made up of a relatively dark central portion (umbra) and a lighter surrounding area (penumbra). Several umbræ may be contained in one mass of penumbra; sometimes the shape of the whole spot is circular, sometimes the outline is complex and irregular. Small spots may be made up entirely of umbra, while in complex groups the penumbral area is widely scattered.

Spots may appear singly, but more often form groups. A common sight is to see two main spots, one lying to the west of the other, with numerous smaller ones near by. In general, the following or easterly spot is the first to decay and vanish. A really large group may contain dozens of separate umbræ, and sometimes the detail is so intricate that it is difficult to photograph and almost impossible to draw.

The average spot lasts for about a week before it disappears, while smaller ones may have a lifetime of less than a day. Occasionally an unusually persistent spot makes its appearance; the record for longevity seems to belong to a spot which was seen from June to December 1943, a total period of nearly 200 days.* The spot was not, of course, under continuous observation for the whole of that period. Since the Sun rotates on its axis, taking rather less than a month to do so, a spot group can be seen moving slowly across the disk as it is carried from east to west. The movement is too gradual to be noticed over short periods, but the shift from one day to the next is very obvious indeed. After a time, the spot will be carried over the western limb, and will not be seen again for a fortnight, after which it will reappear in the east—if, of course, it still exists.

* The spot was followed by the late F. J. Sellers, formerly Director of the Solar Section of the British Astronomical Association, to whom I am indebted for this information.

The numbers of sunspots vary in a semi-regular cycle. Maxima, during which spots are frequent, occur at intervals of about 11 years, with minima in between. During an active period there may be as many as a dozen groups visible at once, while near minimum the whole Sun may be spotless for weeks on end. This cycle was discovered by a German amateur, Heinrich Schwabe, who drew the Sun on every possible day between 1825 and 1843, counting the observable spots and studying their individual characteristics.

Actually, the cycle does not keep strictly to the 11-year period. The interval between successive maxima may be as short as 9 years, or as long as $13\frac{1}{2}$, so that no exact forecasts can be made; moreover, some maxima are more active than others.

The maximum of 1947-48 was very energetic, and the great group of April 1947 was the largest ever seen; at the peak of its development it covered an area of over 7 million square miles. Minimum was reached in 1953-4, to be followed by another maximum in 1957-58. This was the period of the International Geophysical Year, a vast, world-wide co-operative programme involving scientists of more than fifty nations, and during which the Sun was intensively studied. There were large numbers of great spot-groups, some of them comparable with the giants of 1946 and 1947.

Activity died away, as expected, and minimum was reached in 1964, when for a period there were few spot-groups. This lack of spots persisted until well into 1965. During this period another co-operative programme, the International Year of the Quiet Sun, was carried through successfully. During 1966, activity began to increase once more, and some large groups were seen in the late summer. The next maximum may be expected round about 1969.

Regular observation will show that the spots do not appear to move across the Sun in straight lines, except during early June and early December. This is because of the apparent shift in position of the Sun's axis of rotation. The position of the pole for any date can be looked up from the tables contained in a publication such as *The Handbook of the British Astronomical Association*, but the rough diagrams in Fig. 23 may be of help.

During the early part of a cycle, the spots tend to appear some way from the equator, but as the cycle progresses the spots invade lower and lower latitudes. As the cycle draws to its end, and its groups die away, small spots of the new cycle start to appear in high latitudes once more. At minimum, therefore, there are two areas subject to spots: the equatorial, with the last spots of the dying cycle, and the higher-latitude, with the first spots of the new cycle. This behaviour is termed

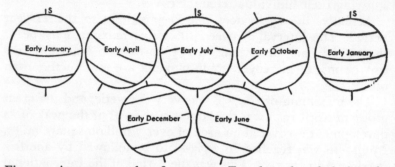

Fig. 23. Apparent paths of sunspots. For the sake of clarity, the apparent shift of the Sun's pole of rotation has been exaggerated.

Spörer's Law, since it was first announced in 1879 by the German astronomer of that name; its cause is very imperfectly understood. It should be added that spots never break out near the Sun's poles of rotation.

Spots are associated with bright irregular patches known as faculæ, from the Latin word meaning "torches". Faculæ appear to lie well above the photosphere, and can be regarded as luminous clouds hanging in the upper regions. They often appear in positions where a spot-group is about to break out, and they persist for some time after the group has disappeared. Consequently, the appearance of faculæ on the Sun's following limb is often an indication that a spot group is coming into view from the far side.

Sunspots possess strong magnetic fields, and emissions from the active regions lead to disturbances of the compass needle, as well as to displays of auroræ, or Polar Lights. It has also been suggested that sunspots affect the weather; the cold winters of 1916, 1927 and 1938 in Britain coincided roughly with maxima, while many people still remember the long "freeze-up" of

1947, when the Sun was so active; the minima of 1921 and 1932 were accompanied by droughts. Nowadays, however, the connection between spots and the weather is regarded as distinctly dubious. The bitter weather in Britain during early 1963 coincided with an approaching solar minimum, so that at best the correlation is unreliable.

Even on the unspotted parts of the Sun, a certain amount of activity is always going on. The photosphere is not at peace; it is covered with "granulations", which are in a state of constant turbulence, and seldom last for long. Generally, a granule has a width of something like 500 miles. The nature of the granules is not definitely known, but the granular structure may well be due to the tops of gas-currents which rise and fall.

Very occasionally, brilliant short-lived patches may be seen over sunspots. The first of these "flares" was seen in 1859 by two amateurs, Carrington and Hodgson, but for many years no more were recorded. Flares visible in ordinary telescopes are in fact so rare that an observer may go through his whole lifetime without seeing one, but modern instruments have shown that the solar flare is a common phenomenon.

The true origin and nature of flares is not yet definitely known, but present evidence seems to point to their being storms in the chromosphere, of an electrical nature, the hydrogen atoms being caused to glow brilliantly by electrical excitation. They spread through large areas of the chromosphere horizontally, i.e. parallel with the solar surface, with amazing rapidity, but there is very little vertical movement; they seem to be confined to the 8,000 or 10,000 miles of the chromospheric depth. They produce marked effects upon the terrestrial compass-needle, as well as helping to cause radio fade-outs and other disturbances.

Features of the upper layers, such as prominences, are best discussed together with eclipses of the Sun, and it will be wise to restrict the present chapter to objects which can be seen with ordinary telescopes.

Observing the Sun is not the simple matter that might be imagined. Even a small telescope can concentrate so much light and heat that an incautious observer who puts his eye to the tube may be blinded. Very great care is necessary at all times; it is only too easy to make a mistake.

Unfortunately, it is possible to buy special dark-lensed "suncaps" which fit over an ordinary eyepiece, and can be used for direct observation. According to some textbooks, it is then safe to turn a 2- or 3-inch refractor directly towards the Sun, and observe in the usual manner. *This is emphatically not the case.* No suncap can give full protection, and in any case there is always a chance that the cap will splinter, so that the eye of the observer will be seriously injured before he has had time to realize what has happened. This warning is not mere alarmism; I know of one amateur who lost the sight of his left eye through an accident of this sort, and the risk is not worth taking, particularly when better observations can be made by indirect means.

There is another danger also. Sometimes the Sun can be seen shining through a layer of thick mist, so that it appears reassuringly dim and gentle. The temptation is then to use a telescope directly, either with or without suncap. Here again there is more than a chance that permanent damage to the eye will result; as soon as the solar radiation is focused, it becomes unsafe. In short, never look straight at the Sun even with binoculars. It is true that some kinds of special eyepieces, known as wedges, are fairly harmless; but the only really sensible way to draw sunspots is by projecting them on to a piece of white card.

Projection is an easy process, since there is plenty of light available. First turn the telescope in the direction of the Sun, "squinting" over the top of the tube and keeping a cover over the object-glass. Then rack out the focus, and remove the cap from the end of the tube. Hold a white card a few inches away from the eyepiece, and move the telescope gently (if necessary) until the image of the Sun appears, after which the disk can be brought to a sharp focus by adjusting the rack and the position of the card (Fig. 24). Any spots and faculæ that happen to be present will be obvious at a glance. A low power is advisable— I have found that for my 3-inch f/12 refractor, ×72 gives good results—though the magnification can be increased for drawings of individual spots on a larger scale.

To make the drawings conveniently standard, it is as well to draw a 6-inch circle on a card and then adjust the distance and focus until the image of the Sun exactly fills it.

If the telescope used is very small, a 4-inch circle may have to suffice.

It is not easy to hold the card steady, move the telescope to follow the Sun, and draw the visible spot-groups at the same time. One would have to be a Briaræus to do so effectively, and the obvious solution is to fit an attachment to the telescope tube which will hold the card at the right distance from the eyepiece. Such an attachment can be built by anyone who is reasonably skilful with his hands, but should any difficulty

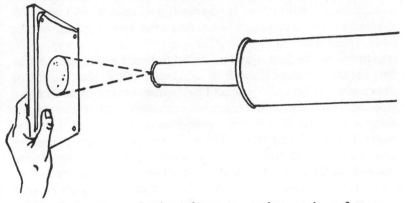

Fig. 24. Simple projection of sunspots, using a 3-in. refractor.

arise the beginner would do well to refer to a paper written in 1938 by F. J. Sellers.*

When the drawing has been finished, the following details should be added: date, time (G.M.T.: never Daylight Saving Time), observer's name, aperture of telescope, and magnification. If any of this information is omitted, the drawing loses most or all of its value.

In general, refractors are to be preferred to reflectors for solar work, and the ideal aperture is from 4 to 5 inches. A 6-inch is larger than is necessary, and extra care must be taken. In the case of a reflector, the mirror should be left unsilvered, which naturally makes the instrument almost useless for any other kind of work. During the many years that I have owned my 12½-inch reflector, I have never turned it towards the Sun, and nor shall I ever do so. My portable 3-inch refractor will

* Journal of the British Astronomical Association, Vol. 49, page 75.

show the spots and faculæ quite well enough, and to use a larger aperture in such a way would be madness.

Spots and faculæ may, of course, be photographed, and no really elaborate equipment is required. Excellent results may be obtained with a modest 3-inch refractor. Outstanding photographs are regularly secured by W. M. Baxter, using the 4-inch refractor at his observatory in Acton; some of these are reproduced on Plate 3, and their quality is obvious.

Such work is decidedly useful. In particular, Baxter has been investigating the so-called Wilson Effect. If a sunspot has a depressed umbra, the "preceding" penumbra will be foreshortened, and will appear narrow as the spot comes over the limb, while when the spot has crossed the disk and is approaching the opposite limb the "front" penumbra will appear to be the broader. In fact, with a circular sunspot which has a depressed umbra, the penumbra closest to the Sun's centre will always seem to be narrower than that on the opposite side of the spot. Baxter has found that the Wilson Effect is often perceptible, so that most sunspots are relatively shallow hollows a few hundred miles deep, but now and then he has studied an unusual spot where the Effect is actually reversed. More research is needed, and amateurs can do valuable work both visually and photographically.

The amateur who is prepared to buy or build a special instrument, such as a spectrohelioscope, will have almost unlimited scope in the field of solar observation. Even a simple spectroscope will show the prominences, and this at least is within the powers of any skilful amateur, as can be seen by reference to the books listed in the Appendix. A Lyot monochromatic filter is even more convenient.

While it would be idle to pretend that the observer who contents himself with drawing sunspots with the aid of a small refractor has much chance of making a valuable discovery, particularly since daily disk photographs are taken at solar observatories, the time spent will not be wasted. Much will be learned; it is fascinating to watch the spots and faculæ as they drift, change and finally die away. Yet we must never forget that we are unworthy to take liberties with the ruler of the Solar System. A cat may look directly at a king, but no telescopic worker must ever look directly at the Sun.

Chapter Six

THE MOON

THE MOON, BY far the nearest of the celestial bodies, is the favourite object of study for the amateur with a small or moderate telescope. Although low magnification may sometimes mean that the eye is dazzled, there is no risk of even slight injury, and observation of the Moon is unaccompanied by many of the troubles that plague the solar enthusiast.

The Moon is almost a quarter of a million miles away. The distance varies somewhat, and the average value is just under 239,000 miles. This may sound a long way, but astronomically it is not far. An aeroplane flying ten times round the terrestrial equator would cover a greater distance than that between the Earth and the Moon. Venus, nearest of the major planets, is always at least a hundred times as remote.

It is therefore not surprising that we can see the Moon really well. A 3-inch refractor will show a surface formation only 3 miles in diameter, while the Meudon 33-inch telescope is capable of revealing a pit only 500 yards across. No other body can be examined in such detail; in fact, low-power binoculars will show the Moon as clearly as Mars or Venus can ever be seen with the most powerful telescope yet made.

Surface markings can be seen with the naked eye, and were noticed in very early times. Who has not heard of the Man in the Moon? He is found in legends from all over the world, and it is certainly true that the bright and dark areas are arranged so that they can sometimes give a vague impression of a human face. When a telescope is used, however, the Old Man vanishes, his features lost in a mass of complex detail.

When Galileo first built his tiny refractor and turned it to the skies, the true nature of the Moon was revealed. Galileo was not the only astronomer to take an interest in the surface features; indeed, it seems that the first telescopic map of the Moon was drawn by an Englishman, Thomas Harriott, while in 1610 we find Sir William Lower making observations from his home in Pembrokeshire and comparing the Moon with a tart that his cook had made: "Here a vaine of bright stuffe, and there of darke, and so confusedlie all over." In 1645 Hevelius, a wealthy Danzig merchant who built an observatory on the roof of his house and installed one of the immensely long, small-aperture refractors, drew a chart of the surface that is still of interest. However, it was not until nearly two centuries later that the first reliable map was published by two Germans, Beer and Mädler, who made their observations with the aid of Beer's 3¾-inch refractor. Other maps have followed, notably that by the late H. P. Wilkins. Photographic atlases have also been produced recently in America by G. P. Kuiper and his colleagues, and in Japan by Miyamoto and Matsui.

However, it must be said that by now all "Earth-based" maps have been superseded by the results from the U. S. Orbiter vehicles. There were five Orbiters altogether, launched between 1966 and 1967, and they sent back photographs of all the visible face of the Moon as well as much of its reverse side. The details shown are quite remarkable, and many of them cannot be seen from Earth even with very large telescopes. It can be said, with truth, that the Moon has at last been properly mapped — but it took rocket vehicles to do so.

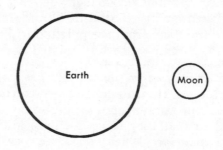

Fig. 25. Comparative sizes of the Earth and Moon.

The Moon is a world quite unlike our Earth. It is much smaller, with a diameter of 2,160 miles—comparable with the width of the Atlantic—and a mass of only 1/81 of that of the Earth (Fig. 25). It is virtually without atmosphere, and is devoid of water, but this was not realized by the seventeenth-century observers. The great dark plains were thought to be seas, and were given suitably romantic Latin names, such as the Mare Imbrium (Sea of Showers), Mare Nectaris (Sea of Nectar), Palus Nebularum (Marsh of Clouds) and Oceanus Procellarum (Ocean of Storms). These names could hardly have been more ill-chosen. Showers, clouds and storms are unknown upon the Moon, and the so-called seas have proved to be mere dry plains.

Once names have been given, it is inconvenient to change them, and the old nomenclature is still used. Perhaps there is a germ of truth in it; we know at least that the "seas" were fluid after the rest of the surface had solidified, since formations lying on or near the former coasts have had their "seaward" walls ruined and broken down. A good example of this is the old crater Fracastorius, shown in Plate XV. p.226 SW

Many of the seas, such as the vast Mare Imbrium* (Plate XVI), are roughly circular in form. Their boundaries are raised, and form veritable mountain ranges, some of which are high by our standards. Typical examples are the Apennines, which form the south-western border of the Mare Imbrium, and which tower to about 15,000 feet; the Alps, lower but still not to be despised; and the Caucasus Mountains, which separate the Mare Imbrium from the neighbouring plain. Earth-type mountain chains are rare, but isolated peaks and clusters of hills are to be numbered in their thousands. The very highest mountains on the Moon attain about 30,000 feet, so that they surpass our Mount Everest, while a modest peak such as Snowdon would appear very humble and inconspicuous if transferred to the lunar surface. The Moon is certainly a rugged world.

The landscape is dominated by the almost innumerable walled formations known collectively as "craters". They are everywhere; they cluster on the bright uplands, breaking and ruining one another, and they are also to be found on the dark seas, the floors and walls of other craters, and on the summits

* It is best to keep to the Latin forms, and this will be the course adopted here. A full list of these names, with their English equivalents, is given in Appendix XII.

of peaks. No part of the Moon is free from them, and they range in size from the colossal formation Bailly, 180 miles in diameter, down to tiny pits at the very limit of visibility. Plate IV (B) shows some vast craters, including the 90-mile Ptolemaeus.

Riccioli, an Italian priest who drew a lunar map in 1651, adopted the convenient but rather dangerous practice of naming craters after famous scientists of the past and present. The system has been followed, and later workers have added to the list, so that by now nearly 700 craters have been given titles. Newton, Flamsteed, Halley, Herschel and Riccioli himself are all there, together with less scientific folk such as Alexander and Julius Caesar; Ptolemy, the "prince of astronomers", has been allotted a 90-mile crater close to the centre of the disk. Smaller craters are lettered. On an outline map such as that given on pages 226-7, it is naturally impossible to give all these names, so that only the larger and the more remarkable formations have been picked out.

Though some of the craters are deep — appropriately enough, the deepest of all is Newton, with walls that attain 29,000 feet above the floor — it would be incorrect to regard them as being like wells. A typical crater has a mountain rampart which rises to only a moderate height above the outer country, but much higher over the deeply depressed interior. The inner walls are often terraced, and the slopes are comparatively gentle, so that a lunar crater resembles a saucer more than a well. This is shown in Fig. 26, in which the 38-mile crater Eratosthenes is drawn in profile. It is shown in Plate XVI; like many other walled formations, it has a central elevation, which looks inconspicuous and yet is really of great altitude. With some craters, the central elevation is a single mountain; inside other formations we see clusters of hills, while central craters are also frequent. On the other hand, there are also formations whose floors appear flat and featureless except under high magnification.

The depths of the craters are measured by studying the lengths of the shadows cast by the walls. This brings us to the question of the "terminator", which is of vital importance to every lunar observer.

The terminator is the boundary between the sunlit and

Fig. 26. Cross-section of the lunar crater Eratosthenes. The diagram is simplified for the sake of clarity; the curvature of the lunar surface is neglected, and the central elevation is shown as a simple peak instead of a complex mass.

night hemispheres. It must not be confused with the true limb, which is merely the edge of the area visible from Earth; this is made clear in Fig. 27, in which the terminator is dotted while the limb is drawn as a continuous line.

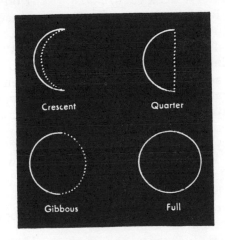

Fig. 27. Limb and terminator. The limb is drawn as a continuous line, while the terminator is dotted.

Though the limb appears more or less smooth, the terminator does not. A peak or a piece of high ground will catch the sunlight in preference to an adjacent depression, and so the whole terminator looks rough and jagged. Sometimes a peak may appear to be completely detached from the rest of the disk, and this effect is shown in Plate XVI.

Since the shadows cast by crater walls and peaks can be measured, and the altitude of the Sun over them is known, the heights of the mountains and the depths of the crater floors can be calculated with accuracy. Towards Full Moon, of course, the shadows almost disappear, and even larger craters become hard to make out unless they have particularly bright walls or particularly dark floors. It is therefore wrong to suppose that Full Moon is the best time for observing; on the contrary, it is the worst, except for special investigations. The most spectacular views are obtained during the crescent half and moderately gibbous stages.

I can cite a personal experience here. I first looked at the Moon through a telescope when I was a boy of eight, and since I knew no better I decided to make a start on the night of Full Moon. I looked up the position of Ptolemaeus, arranged my newly-acquired 3-inch refractor, and tried to find my way about. Naturally, I failed to find Ptolemaeus. When I looked again at the time of half-moon, the crater was still partly filled with shadow, and I could identify it at the first glance.

Though craters near the Moon's centre appear circular, unless they have been distorted by other craters, those near the limb are elliptical. This is purely an effect of foreshortening, due to the fact that the Moon is a globe. Morever, the craters remain in fixed positions; Ptolemaeus is always near the centre of the disk, Mare Imbrium to the north-east, Petavius to the west. This is because the Moon keeps the same hemisphere permanently towards us, an annoying trait due to tidal inter-action with the Earth in past ages.

The Moon revolves round the Earth roughly once a month. It also takes roughly a month to spin on its axis, so that the lunar "day" is 14 times as long as ours, and is followed by a "night" of equal length. This sometimes causes some confusion, but a clumsy experiment will show what is meant. Place a chair in the middle of the room, to represent the Earth, and assume that your head is the Moon, while your nose stands for the almost central crater Ptolemaeus. The back of your head will then represent the "back" of the Moon (Fig. 28). Now walk round the chair, keeping your nose turned toward it all the time. By the time you have completed one circuit, you will have turned round once; your "sidereal period" will have been equal to your "axial rotation", and anyone sitting on the chair will not have seen your back hair at all. This is how the Moon behaves; just as the seated observer failed to see the back of your neck, so the terrestrial observer fails to see the far side of the Moon. Until the flight of the Russian rocket Lunik III, in 1959, nothing was known about these regions.

Fig. 28. A simple demonstration
of the movement of the Moon
round the earth.

However, the situation is not quite so bad as might be thought. Though the Moon spins on its axis at a constant speed, it has a somewhat elliptical path, and this means that its orbital speed varies. When at its closest to the Earth, it moves more quickly than when it is more distant. The axial spin and the position in orbit therefore get periodically out of step, the result being that the Moon seems to rock slowly to and fro about its axis, allowing us to view first the east limb and then the west. On some nights, the Mare Crisium (Plate IV (c)) will appear to be almost touching the limb, while on others it will be well clear. There is also a rocking in a north-south direction, since the Moon's orbit is slightly tilted, and we can thus peer for some distance beyond alternate poles. These rocking motions or "librations" mean that from Earth, we can examine a total of 4/7 of the Moon's surface, while the remaining 3/7 remains permanently hidden.

This may be the moment to say something more about lunar probes, because the first real triumph came with the photography of the long-mysterious "other side of the Moon", carried out by the Russians in 1959. The vehicle on this occasion was Lunik III. Its results seem very outdated now, but it was a magnificent achievement, and at least it cleared up quite a number of problems. As expected, the far side of the Moon proved to be of essentially the same nature as the side we have always known.

Since then there have been the circum-lunar probes, notably the Orbiters, and also the soft-landers. Up to mid-1967 both the Russian Luna vehicles and the U. S. Surveyors had come down gently on the Moon and sent back detailed photographs of scenes which look remarkably like lava-fields. In September 1967 Surveyor No. 5 actually carried out analyses of the Moon's surface materials, and indicated that they were very similar to well-known types of volcanic rocks. And this brings me on to a question which has been the subject of much debate: that of the origin of the Moon's craters.

Many ideas have been proposed, but only two seem to be seriously considered nowadays. Either the Moon's walled formations are volcanic — that is to say, formed by internal action — or else they are due to the impacts of meteorites. It is, and always has been, overwhelmingly probable that both volcanic and impact craters exist, but for the moment we are considering the really large structures of the Clavius, Ptolemaeus or Aristarchus type. During the 1950s, the impact theory was fashionable, and was even claimed to have been proved. People such as myself, who strongly disagreed with this idea and claimed that the craters were much more likely to be volcanic structures of the caldera variety, were subjected to much criticism.

To go fully into the arguments would be beyond the scope of this book; suffice to say that in my view the forms of the craters, their structure and their non-random distribution were alone enough to indicate that most of them (at least, most of the large ones) must be volcanic. Certainly the results of the space-probes seem to support volcanism rather than impact, but of course the last word has by no means been said. It is always dangerous to be dogmatic; and though my ideas have met with strong confirmation during the past few years, I may yet be wrong!

One point has, however, been cleared up. One of the less convincing theories was produced in the 1950s, according to which the Moon's maria were filled with soft dust to a great depth and that to land a space-craft there would be hazardous, as it would simply sink out of sight. The dust theory never made any pretence of fitting the facts, and I am at a loss to explain why it was taken so seriously. Eventually, first Luna 9 and then Surveyor 1 landed on what was obviously a hard surface. Brief attempts were made to explain that the surface was dusty everywhere except just where the probes had come down, but after a while the dust theory quietly died.

Thousands of pictures of the Moon have been taken from the various probes, and many have been published. I have not included any of them here, because we are discussing the research possible with the aid of modest telescopes based on Earth; but it is as well to bear in mind that many of the lunar problems of, say, 1957 are problems no longer, and amateur scope is correspondingly restricted.

Of the minor features, special mention should be made of the clefts or rills, which look like surface cracks and extend in some cases for over 100 miles; an excellent example is the Ariadaeus Cleft, shown on the map in the Appendix (page 226). Close beside it is the Hyginus "cleft", which is basically a crater-chain, and is superbly shown in one of the Orbiter

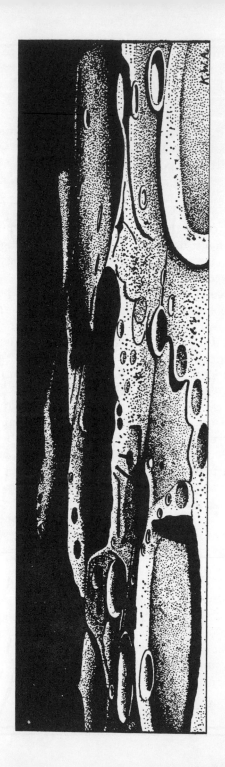

Fig. 29. Line drawings of lunar features. (above) The area of the crater Demonax; K. W. Abineri, January 5, 1950, 11.00–13.30 G.M.A.T.; 8 in. reflector, × 232. (below) Chart of the formation Palitzsch; Patrick Moore; observations made between 1951 and 1954; 12½-in. reflector.

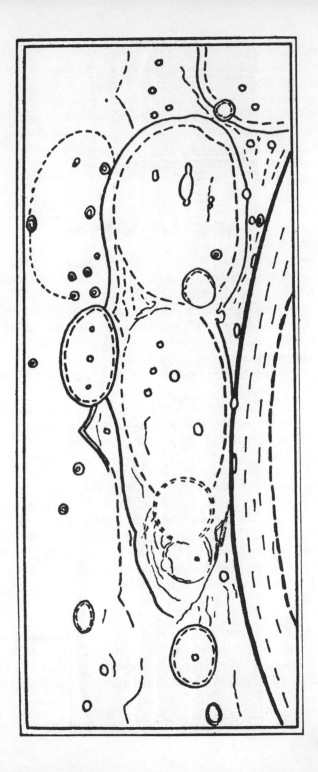

pictures. Look, too, at the cleft inside the giant walled plain Petavius, which runs from near the central mountain across to the wall. Over the whole Moon there are many clefts and cleft-systems accessible to small telescopes, and there are also crater-valleys (Rheita, Reichenbach) and the low swellings known as domes (such as those near Arago). Of course, under high illumination the bright rays associated with certain craters, notably Tycho and Copernicus, dominate the scene. They cast no shadows, and must be surface deposits. There are plenty of minor ray centres (Kepler, Olbers and Anaxagoras, for instance) and there are some craterlets which are surrounded by bright patches; Euclides, near the Riphaean Mountains, is a good example.

Let us repeat that in view of the probe photographs, the lunar work of the amateur astronomer is more restricted than it used to be. Frankly, there is little point now in charting the surface — except for one's personal pleasure, of course. This has been a revolution which observers such as myself have found strange. I spent immense numbers of hours between 1936 and the mid-1960s in making lunar maps, particularly of the limb areas, but they are no longer needed. Orbiter has done the work for us.

Yet it is still essential to learn one's way around the Moon, and I recommend the method that I adopted myself when I began to take a real interest. Using an outline chart, I sketched every named crater and identified it at the telescope. The process took over a year, but it did teach me which crater was which. The main beginners' trap is in using the wrong scale; actually, 15 miles to the inch is quite suitable. The drawings in Fig. 29 show an artistic sketch by K. W. Abineri and a line drawing by myself. Both are out of date in view of the Orbiter coverage, but they do show the two main methods of presentation. Of course, changes in illumination cause remarkable alterations in the aspect of the lunar surface, even from one night to the next, but constant practise will soon bring order into the apparent chaos.

So far, we do not have as much information as might be wished about the depths of some of the lunar craters, and one line of research here is worth following up: that of shadow-estimates. It is assumed that the diameter of the crater is

84

known. The fraction of the east-to-west diameter of the crater is then estimated, in tenths, giving all the information needed for the depth of the crater to be worked out; the calculations are somewhat involved, but depend entirely upon the raw observations.

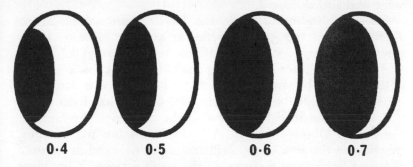

0·4   0·5   0·6   0·7

Fig. 30. Specimen "shadow blanks". It is best to use these for comparison with craters, because otherwise one is apt to be misled. For instance, a glance at the second diagram would probably lead to an estimate of 0·4, whereas the true figure is 0·5; if you measure it, you will see that the edge of the shadow lies exactly half-way between the walls, measured in a west-east direction (across the page from left to right). A whole series can be done between 0·3 and 0·8; it is not safe to go outside these limits, and the best estimates are made between 0·4 and 0·7.

I have found that the best method is to prepare a series of disks, as shown in Fig. 30, and compare the disks with the craters concerned. The accuracy of the method is quite surprising, and it has been followed with success by the Lunar Section of the British Astronomical Association. A telescope of some aperture is needed, since the method is useful only for small features; but a 6-inch reflector is adequate. It is clearly important to note the time of observation, to the nearest minute, because the calculations depend upon knowing the angle at which the Sun's rays are striking the walls when the estimate is made.

Another field of research concerns possible variations on the Moon's surface. Here, too, opinions have changed of late. It used to be officially laid down that the Moon is changeless; admittedly, there was one alleged case of alteration—in the formation Linné, on the Mare Serenitátis, which was drawn as a crater by all observers before 1843 and has been recorded as a small pit, surrounded by a white patch, since 1866—but the evidence was most uncertain. The periodical red patches and local obscurations recorded by amateurs were dismissed as being due to errors of observation, even though some of them were well authenticated.

Then, in November 1958, the Russian astronomer N. A. Kozirev, using the 50-inch reflector at the Crimean Astrophysical Observatory, recorded some temporary red colour inside the larger crater Alphonsus, one of the best-known features of the Moon. Though the cause and interpretation of Kozirev's observation could be questioned, the observation itself could not, and it provided food for thought. In 1963, observers at the Lowell Observatory, in Arizona, saw more red patches, this time in the area of the brilliant crater Aristarchus—one of the most fascinating features on the lunar surface; it lies close to the imposing valley which extends from the adjacent crater, Herodotus. In this case, too, the observations could not be questioned.

I may perhaps be allowed to cite a third case of red colour, because I was concerned in it myself. On April 30 1966, the British amateur astronomer P. K. Sartory, from his observatory in Surrey, detected redness close to Gassendi, a large crater on the borders of the Mare Humorum. The Lunar Section of the British Astronomical Association had established an organization to study this sort of phenomenon, and the colour patches were fully confirmed by independent observers—by P. Ringsdore in Surrey, and by T. J. C. A. Moseley and myself at Armagh. We did not compare notes until later, and the agreement was so good that no reasonable doubt remains.

Sartory's original observation was obtained with a device known as a Moon-Blink. This takes the form of a rotating wheel, with red and blue colour filters, placed just on the object-glass side of the eyepiece (or the mirror side, with a reflector). A red patch will show up as a dark feature when observed through a blue filter, but will be masked with a red filter. Rotating the wheel, one observes first through the red, then through the blue filter in quick succession, so that any red patch will show up as a "blinking spot".

Subsequently, further colour phenomena have been detected

in the Gassendi region and in other parts of the Moon, and the Moon-blink programme has become really important. Sometimes the redness can be seen directly; at others, it is betrayed only with the filters—but it now looks as though these "transient lunar phenomena", or T.L.P.'s for short, are less uncommon than has been believed. There are also some local areas of the Moon which show weak permanent blinks, indicating a faint redness present all the time and quite different in nature from a true T.L.P.

It is not hard to make a Moon-blink device, and it can be used with a modest telescope, though I would not be happy with anything much smaller than an 8-inch. Yet here, above all, it is vital to avoid jumping to conclusions, and the greatest care is needed. It is only too easy to "see" what one half-expects to see, and a bad observation is worse than useless—it is actively misleading.

I do not propose here to discuss the cause of these transient phenomena, except to say that it seems as though emissions from the lunar ground seems to be the most likely cause. But the work carried out in this direction seems to show that the Moon is not quite inert—and most of this work has been handled by amateurs.

Excellent amateur photographs of the Moon can be taken, some of which are given in Plates IV and V. They cannot compete with those of the Orbiters, but they are interesting to take and they are valuable inasmuch as they can provide photographic outlines for craters under study.

What, then, is the future of lunar observation?

It is difficult to tell. The Moon is so very different from the Earth; it is virtually without atmosphere, and therefore it is presumably without life. It used to be thought that certain dark patches in some craters, such as Aristillus, might be due to lowly organisms, and the same was suggested for the strange radial dark bands inside Aristarchus and others of its kind; but this attractive idea has long since been given up. Most of the most important lunar puzzles will be solved soon by new rocket probes, as many problems have been solved already. Yet even when the Moon has been reached, there will still be endless pleasure in looking at the craters, the mountains and the valleys, learning how to recognize them and watching how their shadows shift and change as the Sun rises over them.

Chapter Seven

OCCULTATIONS AND ECLIPSES

So far as the Solar System is concerned, there are long periods in which the observer has to content himself with purely routine work. This may well be followed by a number of interesting phenomena which occur in quick succession. There is perhaps a violent outbreak of sunspots, or a favourable opposition of Mars; a bright comet may make a dramatic and unexpected appearance. Also to be considered are occultations and eclipses, which have been described as the celestial equivalents of hide-and-seek.

The Moon is much the closest body in the sky, and so moves across the starry background at a relatively high speed. Sometimes it must, of course, pass in front of a star and hide it. These "occultations" are common enough, but are not so numerous as might be thought. People tend to over-estimate the size of the Moon in the heavens, and artists will usually draw it as large as a dinner-plate, whereas the apparent diameter is the same as that of a quarter held 9 feet away from the eye.

Consequently, the Moon does not pass in front of several stars per hour. During 1967, for instance, as seen from Britain there were no occultations of stars bright enough to be shown in the star-maps in the Appendix. The Moon, like all the planets, keeps to the Zodiac, so that only stars close to the ecliptic can be occulted; the brightest of these are Antares, Aldebaran, Spica and Regulus. Occultations of planets can also occur at times.

A planet shows a definite disk, so that it takes some seconds for the oncoming limb of the Moon to pass right over it. A star, however, appears as a tiny point of light, and the disappearance is virtually instantaneous. The star shines steadily until the moment of occultation, and then seems to snap out like a candle-flame in the wind. One moment it is there, the next it has gone. This is one proof that the Moon has little or no atmosphere, since a blanket of air around the limb would make the star flicker and fade for some seconds before vanishing.

Seen in a telescope, an occultation is a fascinating sight. The star seems to creep up to the Moon's limb, though actually the Moon's own motion is responsible, and the inexperienced observer is bound to feel that the star hangs close to the limb for a long time. Then the brilliant point of light will "softly and suddenly vanish away", like the hunter of the Snark, and a watcher who blinks his eyes at the wrong moment may easily miss the disappearance. The emersion, at the far limb, is equally abrupt.

Occultations are more important than one might think. They can be predicted, and *The Handbook of the British Astronomical Association* gives a list for each year, but the predictions may not be absolutely correct, owing to the fact that the Moon's apparent path in the sky is not known with complete precision. The star positions are much more certain, and so an occultation enables astronomers to correct their lunar tables. If the disappearance is timed accurately, it gives the actual position of the Moon's limb at that moment.

This is work that the amateur can do, but, like all other observations, it must be carried out with extreme care. I once had a report from an observer who said that the star Omega Leonis "was occulted at about twenty minutes past ten". This sort of thing is useless. The occultation must be timed to an accuracy well within one second of time if it is to be of any value whatsoever. A really good stop-watch is essential.

When an occultation report is drawn up, the following data should be added: name or number of star, time of occultation, latitude and longitude of observing station, height above mean sea level of observing station, atmospheric conditions, and any peculiar appearance seen. Occasionally, a star is hidden by a lunar peak on the limb and then reappears briefly in the adjacent valley before vanishing once more, so that it seems to wink. This cannot be predicted, since we do not know either the position of the Moon or the contour of the limb with sufficient accuracy.

Now and then, unexpected occultations take place—

89

unexpected in the sense that they are not listed in the available tables. During one lunar eclipse, I witnessed an occultation of the planet Uranus that I certainly did not anticipate. The tables at my disposal made no mention of it, though the occultation was of course known to those who had made calculations beforehand.

Patience and practice are vital in all occultation work, but the time taken will be amply repaid by the fact that valuable observations can be made. A small telescope will prove perfectly suitable, and can well be mounted on an altazimuth provided that the stand is steady.

Planets, too, can occult stars. The most valuable of the planetary occultations are those due to Venus, because the flickering and fading of the star before disappearance gives a clue as to the height of the atmosphere surrounding that mysterious world. On July 7, 1959, for instance, Venus occulted Regulus; the phenomenon took place in the early afternoon, and observations made by Henry Brinton and myself, using Brinton's $12\frac{1}{2}$-inch reflector, showed a perceptible dimming lasting for almost one second. Unfortunately, occultations by planets are comparatively rare, but they are very well worth observing.

It should be added that there have been cases of one planet hiding another; for instance, Venus occulted Mars in 1590 and Mercury in 1737. These phenomena are of course very rare, and few observers will be lucky enough to see one during the course of a lifetime. Recently, it has been pointed out that even the faint, far-off planet Pluto is capable of causing occultations; if such a phenomenon could be watched, the time taken for the star to pass behind the planet would give some indication of Pluto's apparent diameter, which is still uncertain. Unfortunately a large telescope would be necessary.

Occultations are interesting, but eclipses are genuinely spectacular, and are bound to excite the interest even of the non-scientist. A solar eclipse is merely an occultation of the Sun by the Moon, but a lunar eclipse is very different, since the Moon is not hidden by any solid body, but passes into the cone of shadow cast by the Earth.

The principle is shown in Fig. 31. The Moon has no light of its own, so that when it enters the Earth's shadow it turns a dim, sometimes coppery colour. The main cone, shaded in the diagram, is known as the umbra,* while to either side of it is the penumbra, caused by the fact that the Sun is a disk and not a sharp point of light. The diagram is not, of course, to scale, but it does serve to show what happens.

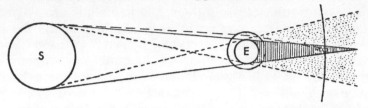

Fig. 31. Theory of a lunar eclipse. S—Sun; E—Earth; m—the position of the Moon at mid-totality. The diagram is not to scale.

If the Moon passes right into the umbra, the eclipse is total. Every scrap of direct sunlight is cut off, but some of the Sun's rays will still reach the Moon, as they are bent or "refracted" on to it by the Earth's mantle of atmosphere, as is shown by the dashed line in the diagram. The result is that instead of vanishing completely, the Moon can usually be found without difficulty even with the naked eye. However, all eclipses are not equally dark. In 1761 the Moon disappeared so completely that it could not be seen at all, whereas in 1848 the totally eclipsed disk still shone so brightly that many people refused to believe that an eclipse was in progress. These variations have little to do with the Moon itself, but are due mainly to the changing conditions in our own atmosphere. It seems for instance that dust in the upper air in 1950, while vast forest fires were raging in Canada, caused the September eclipse of that year to be rather darker than usual. Also very dark was the eclipse of June 25 1964, when, from Sussex, I lost the Moon during totality even with my 12½-inch reflector, though conditions were not ideal. The cause on this occasion was volcanic dust which had been sent into the upper atmosphere by an earlier eruption in the East Indies. By the eclipse of the following December, much of this dust had settled, and the eclipse was lighter, though still rather dark by normal eclipse standards.

* As used here, the terms "umbra" and "penumbra" have of course no connection with the umbra and penumbra of sunspots.

If the Moon does not enter wholly into the umbra, the eclipse is partial, while at other times it is merely penumbral. Penumbral eclipses will not be noticed except by the attentive watcher, since the dimming is too slight to be conspicuous.

Two things are clear from the diagram. First, a lunar eclipse must be visible from one complete hemisphere of the Earth, provided that clouds do not conceal it, and if it is total anywhere it must be total everywhere. Secondly, an eclipse can happen only at Full Moon. If the Moon passes through the centre of the umbra, it may remain totally immersed for over an hour, while the partial phases can extend over four hours.

Lunar eclipses are so obvious that they must have been known from very early times. Were the Moon's orbit not tilted across the ecliptic, a total eclipse would happen at every Full Moon, but the inclination of the Moon's path is enough to prevent this from happening. Imagine two hoops hinged along a diameter, and crossing each other (Fig. 32). The points at which the two hoops cross are called the "nodes", and unless Full Moon occurs very near a node the Moon will miss the shadow altogether, so that no eclipse will occur.

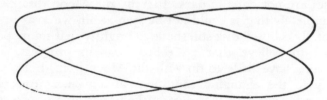

Fig. 32. Two hoops, demonstrating the tilt of the
Moon's orbit.

It so happens that the Sun, Moon and node return to the same relative positions after a period of 18 years $10\frac{1}{4}$ days, and so any particular eclipse will be followed by another eclipse 18 years $10\frac{1}{4}$ days later. This is the so-called Saros Period, and was used by the Greek astronomers to make eclipse predictions. The Saros is not exact, but the method is accurate enough to be workable. A list of future eclipses is given in Appendix IX.

It is interesting to note the different colours seen on the eclipsed Moon, and to see whether the eclipse is bright or dark, but the most important work to be done is in connection with the Moon itself. Since there is virtually no atmosphere to protect the surface, and since the layer of ash is bad at retaining heat, the temperature drops suddenly on the lunar surface as the eclipse begins. The fall may amount to 100 degrees in the course of an hour, and it has been suggested that the abrupt cooling may produce observable effects. Suitable areas for investigation are dark-floored craters such as Plato and Grimaldi, and very bright features such as Aristarchus. There is one formation, Linné on the Mare Serenitatis (Plate XV) which as noted in Chapter VI, was described by the early nineteenth-century observers as a deep crater, but is now a white spot with a tiny central pit, so that it has been suspected of definite change. Various authorities have stated that the white area becomes more prominent during and after the period of cold caused by the eclipse, and if this could be proved it would be of the highest interest, though my own observations have been negative and I am frankly sceptical.

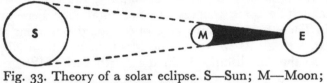

Fig. 33. Theory of a solar eclipse. S—Sun; M—Moon; E—Earth. The diagram is not to scale.

Another field of research concerns the possibility of shooting-stars in the Moon's atmosphere. It was formerly believed that the Moon might still have a mantle with a maximum density of 1/10,000 of ours, which would be quite sufficient to burn up meteors that might venture into it. A meteor in the lunar atmosphere would appear as a short, slow, bright streak, probably ending in a minute flash as the remnants of the falling body struck the ground. Streaks of this sort have been reported from time to time, but since it now seems that the density of the Moon's atmosphere is much less than 1/10,000 the theory of lunar meteors must be regarded as dubious. However, such phenomena are well worth looking for, and an eclipse is the best possible time, because the glare from the surface is then removed.

Turning now to eclipses of the Sun, we find that the principles involved are very different. We are back to the "occultation" idea, since a solar eclipse is caused simply by the Moon passing between the Sun and the Earth.

The Moon is far smaller than the Sun, but it is also so much nearer that in our skies it looks almost exactly the same size. When the three bodies move into a straight line, with the Moon in the middle, the shadow cast by the Moon just touches the Earth's surface, and for a few minutes the bright solar disk is blotted out by the dark, and therefore invisible, body of the Moon (Fig. 33). The width of the completely shadowed area of the Earth is 167 miles at best, and so a total solar eclipse will not be seen over a complete hemisphere; for instance, the eclipse of June 30, 1954 was total in parts of Norway and Sweden, but not in England, where it was partial. This means that although total eclipses are not particularly rare, they are very infrequent at any particular spot on the Earth's surface. The last to be visible in any part of England was that of 1927, while the next will not occur until 1999.

To either side of the track of totality, the eclipse will be partial, while some eclipses are not total anywhere on the Earth. There is also a third kind of eclipse, the annular (Latin *annulus*, a ring). As we know, the Moon moves in an elliptical path, so that its distance from the Earth varies. When at its most remote, it appears smaller than the Sun in the sky, and so cannot cover the whole of the solar disk. When the three bodies line up under these conditions, a bright ring of the Sun is left showing round the dark mass of the Moon.

Obviously, a solar eclipse can happen only at New Moon, and then only if New Moon occurs near a node. The Saros period is valid, as for lunar eclipses, but the rough and ready method of forecasting is less accurate. For instance, the "return" of the 1927 eclipse took place in 1945, but in this year the band of totality lay further to the north, and missed England altogether, so that only a partial eclipse was seen in our country.

Partial and annular eclipses are spectacular, but do not give much scope for useful work. Remember, too, that even when most of the Sun is hidden it is still unsafe to use direct vision either with binoculars or with a telescope. The slightest sliver of sunlight remaining is enough to damage the eye in a matter of seconds, and this was stressed during the large partial eclipses visible in Britain in 1954 and 1961, when there were many cases of people being injured in this way.

A total eclipse is among the grandest of Nature's displays. As the Moon sweeps on, the light fades, until at the instant the last of the disk is blotted out the atmosphere of the Sun leaps into view. There are the magnificent red prominences; there is the glorious chromosphere, and there too is the "pearly crown" or corona, a superb glow surrounding the eclipsed Sun, sometimes fairly regular in outline and sometimes sending out streamers across the heavens. It is a pity that the spectacle is so brief. No total eclipse can last for more than about 8 minutes, and most are much shorter, so that astronomers are ready to travel to remote parts of the world in order to make the most of their limited opportunities. This enthusiasm is not merely for the beauty of the sight; there are many investigations that cannot be made except during the period of totality and the few seconds before and after. In fact, serious workers are so busy that they have no time to stop and admire what is going on.

The prominences are visible to the naked eye only during totality, but with special instruments they can be seen at any time. They are made up of incandescent gas, and are of tremendous size; the length of an average prominence has been given as 125,000 miles. Many are associated with sunspots, and prominences too are affected by the 11-year solar cycle.

"Quiescent" prominences are relatively calm, as their name suggests, and they may last for several months before either breaking up gradually or being violently disrupted. Active prominences may be likened to tall tree-like structures, from the tops of which glowing streamers flow out horizontally and then curve downwards towards the bright surface of the Sun. Some of these active prominences are truly eruptive, and it has been known for the blown-off material to move at over 400 miles per second.

Some of the prominences seen during total eclipses have been curiously shaped. One, seen during an eclipse early in the present century, bore a marked resemblance to an anteater! But since prominence study is not now limited to the period of totality, a great deal of information has been gained as to their behaviour. French and American astronomers have even taken moving pictures of them, and these films are dramatic in every sense of the word.

The pearly corona forms the Sun's outer atmosphere. It is much more extended than the chromosphere, and even the best instruments of to-day can do no more than show the most brilliant parts of it except during an eclipse, so that we still have to rely upon the natural screen provided by the Moon. The corona is made up of very tenuous gas, and stretches outwards from the Sun for many millions of miles, although owing to its low density and indefinite boundary it is not possible to give an exact figure for its "depth".

Although Britain has to wait until 1999 for its next total eclipse, other parts of the world are more favoured. From the list given in Appendix IX, it will be seen that so far as Europe is concerned the next opportunities will be in 1968 (North Russia) and 1973 (South Greece). Those who are prepared to make the journey to more remote areas will be amply rewarded.

It may be of interest to say something about one of the recent European total eclipses, that of June 30, 1954. It was just total off the coast of North Scotland, so that a large partial eclipse was seen in England, and caused general interest even among people not usually astronomically-minded. The main track crossed Scandinavia, where many astronomers gathered. The combined Royal Astronomical Society and British Astronomical Association party, of which I was a member, made its headquarters at the little Swedish town of Lysekil, along the coast from Göteborg, since weather conditions in West Sweden were expected to be rather better than in Norway (as did indeed prove to be the case). Our arrival in Lysekil coincided with the Midsummer Festival. It also coincided with a burst of torrential rain.

On June 30, most observers collected their equipment and drove to Strömstad, in the exact centre of the track, almost on the Norwegian frontier. The site selected was a hill overlooking Strömstad itself, and by noon it was littered with equipment of all kinds: telescopes, spectroscopes, thermometers, cameras, and even a large roll of white paper that I had spread out in the hope of recording shadow bands. These shadow bands are curious wavy lines which appear just before totality. They are due to atmospheric effects, but have never been properly photographed, and the opportunity seemed too good to be missed.* The early stages of the eclipse were well seen. Five minutes before totality, everything became strangely still, and over the hills we could see the approaching area of gloom. Then suddenly, totality was upon us. The corona flashed into view round the dark body of the Moon, a glorious aureole of light that made one realize the inadequacy of a mere photograph. The sky was fairly clear; and although a thin layer of upper cloud persisted, only those with the experience of former eclipses could appreciate that we were not seeing the phenomenon in its full splendour.

It was not really dark. Considerable light remained, and of the stars and planets only Venus shone forth. Yet the eclipsed Sun was a superb sight indeed, with brilliant inner corona and conspicuous prominences. The two and a half minutes of totality seemed to race by. Then a magnificent red-gold flash heralded the reappearance of the chromosphere; there was the momentary effect of a "diamond ring", and then totality was over, with the corona and prominences lost in the glare and the world waking once more to its everyday life. In a few minutes, it was almost as though the eclipse had never been.

The last total eclipse visible in Europe was that of February 15, 1961. The track of totality extended across South France, North Italy, Jugoslavia and the southern U.S.S.R., in the course of which it covered three major observatories—St. Michel in France, Arcetri in Italy, and the Crimean Astrophysical Observatory. Much valuable information was obtained. Again I have personal recollections, since we decided to show the eclipse on television, and I was dispatched to the top of Mount Jastrebac in Jugoslavia to broadcast the commentary from there. Amazingly, the programme was successful, and millions of people all over Europe were able to see totality on their television screens. I also remember the difficulties I had in communicating with the Jugoslav camera director. Eventually I talked French to a Belgian astronomer, who relayed it in German to the Jugoslav. The method was

---

* Unfortunately, conditions at Strömstad were not suitable for the appearance of shadow bands, and none were seen. One eminent astronomer went to the trouble of photographing my apparatus, remarking dryly that although he might live to see another total eclipse, he would never again see so peculiar an arrangement!

slightly cumbersome, but fortunately it worked. Clouds hid the early part of totality, but by good fortune they cleared in time for us to see the end stages and the magnificent "diamond ring".

Since total eclipses are so elusive, the opportunity to watch one should never be missed, even if no useful work is to be attempted. We must be thankful that we are privileged to see the spectacle at all. The fact that the Sun and Moon appear so nearly equal in size can be due only to chance; were the Moon a little smaller, or a little more distant, the solar corona might still remain unknown.

Chapter Eight

AURORÆ AND THE ZODIACAL LIGHT

IT IS IMPOSSIBLE to separate one science from another. Even astronomy is no longer "on its own", as it used to be. It is bound up closely with chemistry and physics, and it is also linked with weather study, or meteorology, by the phenomenon known as the Aurora Polaris, or Polar Light.

Auroræ have been known from very early times, and are so common in high latitudes that a night in North Norway or Antarctica would seem drab without them. In England they are less frequent, though displays are seen on an average at least ten times a year, while in the tropics they are rare. They are not unknown; there is a famous story of how the cohorts of the Roman emperor Tiberius once rushed northwards to the help of the people of Ostia because of a red glow in the sky that they took for a tremendous fire, but which proved to be merely an aurora. However, there can be no doubt that observers in the far north and south have the best views.

Auroræ occur in the upper atmosphere, at heights ranging from over 600 miles down to as low as 60. Sometimes the lights take the form of regular patterns, while at others they shift and change rapidly, often showing brilliant colours and providing a spectacle that is second only to the glory of a total solar eclipse. One of the greatest displays of modern times took place on January 25, 1938, when all Britain witnessed the spectacle. From Cornwall "the whole of the western sky was lit with a vivid red glow like a huge neon sign; gradually shafts of white light were intermingled with the redness, changing quickly to an uncanny grey light and then to a brilliant silver, while green patches appeared here and there". From Sussex the dominating colour was red, though during the course of my own observations I recorded many other hues as well. The aurora was brilliant and widespread enough to cause interest and even alarm over the whole country, and it was seen from places as far south as Vienna.

Since meteorology is the science of the atmosphere, and

auroræ are definitely atmospheric phenomena, one might at first think that they are outside the scope of the astronomer. Yet the cause of auroræ is to be found not on Earth, but in the Sun. Certain active regions of the Sun's disk send out electrified particles, and it is these particles which enter our air and give rise to the glow, though the process is not completely understood—and is certainly associated with the so-called Van Allen zones, which are belts of charged particles surrounding the Earth. The Van Allen belts were quite unsuspected until 1958, when they were detected by instruments carried aboard Explorer I, the first successful American artificial satellite.

Active regions on the Sun are often associated with spots, so that auroræ are most frequent at or near spot maximum. Moreover, a major flare occurring near the centre of the solar disk is often followed a day later by a bright aurora. Since the particles must therefore cover the 93-million-mile gap in about 24 hours, this delay indicates a speed of about 1,000 miles per second.

The fact that auroræ occur mainly in high latitudes is due not to the geographical poles, but to the poles of magnetism. Since the particles which produce the glow are electrified, they must be subject to magnetic attraction, and they are often accompanied by "magnetic storms", radio fade-outs, and other disturbances of like nature. There have even been accounts of hissing sounds heard plainly during the course of a display, though these noises are difficult to explain.

Scientifically, auroræ are important not only because of their link with the Sun, but because they provide information about the upper air. It is therefore useful to observe them whenever possible, and to make estimates of their positions against the starry background, so that their heights may be worked out. The main work here has been done by Norwegian scientists, led by Professor Carl Størmer of Oslo; but amateurs can play a major rôle, and in recent years a full-scale survey has been organized by the British Astronomical Association's Auroræ and Zodiacal Light Section, under the directorship of J. Paton. Observers taking part are asked to fill in forms telling of the presence or absence of aurora, coupled with notes of any displays that may be seen. Negative reports are not without value, and may in fact be of great help.

So far as England is concerned, observing auroræ is made difficult by the innumerable artificial lights. My own experience is a case in point. From my old home in Sussex I used to be a member of the B.A.A. Auroral Survey, but the building of a "new town" in the most inconvenient direction caused a perpetual glare in the sky, so that my work was brought to an abrupt end. Though occasional auroræ become so striking that they cannot possibly be missed, most displays take the form of inconspicuous glows low down in the north or north-west, so that the street lamps of a town are enough to obscure them. Scottish workers are better placed, both because the lights are more scattered and because auroræ are much more frequent.

The lights are so varied that to describe all the forms would need many pages. One never knows quite what an aurora is going to do next, but a great display often begins as a glow on the horizon, rising slowly to become an arc. After a while, the bottom of the arc brightens, sending forth streamers, after which the arc itself loses its regular shape and develops folds like those of a radiant curtain. If the streamers extend beyond the zenith, or overhead point of the sky, they converge in a patch to form a corona (not, of course, to be confused with the corona which surrounds the Sun). Finally the display sends waves of light flaming up from the horizon towards the zenith, after which the light dies gradually away. The whole pheno-menon may extend over hours.

For observing auroræ, by far the best instrument is the naked eye, coupled with a red torch and a reliable watch. Binoculars are of little help, and telescopes absolutely useless. Points to note are the bearing of the centre of the display, reckoned in degrees (o to 360) from north round by east; the type and prominence of the aurora; the various forms seen, such as arcs, curtains, draperies and flaming surges; colours, and duration. Times should be taken at least to the nearest minute. There is obvious scope for the photographer, and spectroscopic work is of great interest, but simple naked-eye observation is not to be despised.

Though auroræ are so spectacular, they are not the only lights seen in the heavens. The sky itself seems to shine with a feeble radiance known as the airglow, and sometimes a cone of light can be seen after dusk or before dawn, extending upwards from the hidden Sun and tapering toward the zenith. Since it extends along the Zodiac, this cone is known as the Zodiacal Light. It can be quite prominent when seen from countries where the air is clear and dust-free, but from Britain it is always hard to see. The Zodiacal Band, a faint, parallel-sided extension of the cone, may extend right across the sky to the far horizon, though it is so dim that it is seldom to be observed at all except from the tropics.

Unlike the aurora, the Zodiacal Light originates well beyond the top of the air. It is thought to be due to light reflected from a layer of thinly-spread matter extending from the Sun out beyond the orbit of the Earth, rather like a tremendous plate. The layer cannot be broad, as is shown by the fact that the Light is never seen except close to the ecliptic. The best times for observation are late evenings in March and early mornings in September, because at these times the ecliptic is most nearly perpendicular to the horizon, and the Light is thus higher in the sky.

Since the Zodiacal Light is so faint, its intensity is not easy to estimate. The best way to measure it is to compare it with a definite area of the Milky Way, and the width of the base, in degrees, should also be noted. Though the Light is predominately white, a pinkish or at least warmish glow has been reported in the lower parts, and should be looked for.

Last and most elusive of these glows is the Gegenschein, which is a faint hazy patch of light always seen exactly opposite the Sun in the sky. It appears at its most conspicuous in September, when it looks like a round luminous patch about forty times the apparent width of the Moon, but it is extremely hard to see, and even a distant lamp is enough to hide it. From England I have looked for it frequently, but have seen it only once—and then not with certainty. Many theories have been advanced to explain it, but at present we have to admit that it remains something of a mystery. The German name is generally used, though some prefer the English term of Counterglow.

For all these observations, one thing should be borne in mind: Never begin work before you have made your eyes thoroughly accustomed to the dark. To come outdoors from a brilliantly-lit room and expect to see an auroral glow or the Zodiacal Light straight away is fruitless, and it is usually necessary to walk about for at least a quarter of an hour before starting your programme, though the exact period is bound to vary with different people. For recording observations, a torch with a red bulb is the ideal, since an ordinary white light will dazzle you sufficiently to ruin the sensitivity of your eyes for some minutes afterwards.

Here again, then, the amateur has a part to play. There is no need to wait years for a great aurora; studying the fainter lights and glows is a fascinating hobby, and it is a pity that city dwellers never have a chance to see the ghostly beauty of the Zodiacal Light.

Chapter Nine

THE NEARER PLANETS

LESS THAN TWO centuries ago, it was thought quite possible that the Moon might be inhabited. Sir William Herschel, the greatest astronomer of his day, regarded the existence of Moon-Men as "an absolute certainty", and we cannot blame him. After all, the Earth is an ordinary planet, so why should it be the only world to harbour life?

As knowledge grew, and the nature of the Moon became more and more clear, the "other men" faded away into the realms of fantasy; but Mars and Venus, particularly Mars, were obviously more promising. And this is one real reason for the ever-present interest in the planets: can they, too, be peopled by beings at least as advanced as ourselves?

To-day the answer seems to be in the negative. Rocket research indicates that Venus is hopelessly hostile; on the other hand Mars does seem to support vegetation, so that it would be hasty and unwise to dismiss the planets with a mere sweep of the hand as dead, uninteresting globes. Professional astronomers, busy with greater problems, are justified in spending all their time on stellar research, but from the amateur's point of view no galaxy or variable star can be more intriguing than a Martian landscape.

Even a modest 3-inch refractor will show markings on some of the planets, but it is difficult to set a limit for the smallest aperture which can be used for serious work. Mars, for instance, needs a larger telescope than Jupiter. Each planet has its own characteristics, and it is best to consider them one by one.

The four members of the inner group—Mercury, Venus, the Earth and Mars—are solid, rocky bodies. They are comparable in size, and all have atmospheres of a kind (though there are doubts in the case of Mercury). These are the only common factors. Otherwise, they are as different as they can be.

Mercury, whirling round the Sun at an average distance of only 36 million miles, is never easy to observe. It always lies somewhere near the Sun's line of sight, and we can well understand why it was named after the elusive, fleet-footed Messenger of the Gods. Moreover it is not much larger than the Moon, and is more than 200 times as distant, so that ordinary telescopes will show little except a pinkish disk with its characteristic phase.

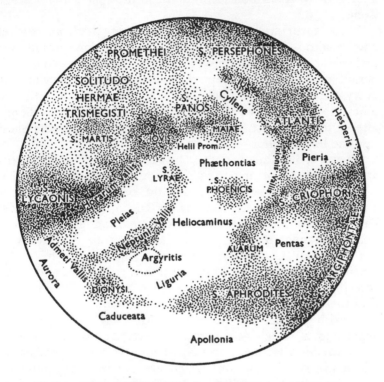

Fig. 34. Map of Mercury, showing the main features, Antoniadi's nomenclature is used.

The rotation period of Mercury has provided a great deal of discussion lately. It used to be thought that the "day" must be equal to the "year"—88 Earth-days in each case. If so, then Mercury would always keep the same face toward the Sun, just as the Moon does with respect to the Earth. Over part of the surface there would be perpetual daylight, with a surface temperature exceeding 700 degrees Fahrenheit; over another part there would be everlasting night, so that the surface would be colder even than remote Pluto. Between these two extremes there would be a "Twilight Zone" over which the Sun would rise and set. Mercury's orbit is not circular, and so its velocity varies between $36\frac{1}{2}$ miles per second at perihelion and only 24 miles per second at aphelion. This would result in effects analogous to the Moon's librations, so producing the Twilight Zone.

Then, surprisingly, American scientists used an entirely new method—radar—to measure the rotation period, and arrived at a value of between 58 and 59 days. This is the period now officially accepted. If it is correct, it means that every part of Mercury will be in sunshine at some time or other, so that our whole idea about the conditions there is altered to some extent even though it does not make the planet any the less hostile.

There is still some dispute as to whether Mercury has a true atmosphere. According to A. Dollfus, in 1950, there may be a thin atmosphere giving a ground barometric pressure of 1 mm. or so; in 1963 N. Kozirev, in Russia, announced that he had detected a very rarefied hydrogen mantle perhaps of a semi-permanent nature; and in January 1964 V. Morov, at the Crimean Observatory, reported indications of carbon dioxide. However, the whole question is still open, and in any case the atmosphere may be regarded as negligible. It seems that there is no chance of any life on Mercury.

Darkish patches can be seen on Mercury, and twenty-five years ago a Greek astronomer, E. M. Antoniadi, used the great Meudon refractor to map them. He gave them names, which are still in use and which are not inappropriate; one area, for instance, is called the "Solitudo Hermæ Trismegisti", or "Wilderness of Hermes the Thrice Greatest". The markings are permanent, as is only to be expected, but we know little about their nature. Mercury may be flat, or it may be mountainous; it may have a smooth surface, or it may be covered with craters like those of the Moon. We simply cannot see it well enough to find out. And I must admit that I doubt whether Antoniadi's map is even approximately accurate, though intensive work with a giant telescope would be needed to produce anything better.

I have glimpsed a few patches with a 6-inch refractor, but no serious work can be done with amateur-owned telescopes. This does not mean that there is no point in looking for Mercury. It is always satisfying to see the strange little world glittering shyly in the late evening or early morning, and on an average it can be seen with the naked eye at least a dozen times each year. There is a legend that Copernicus never saw Mercury in his life, owing to mists rising from the river near his home, but the story is probably untrue.

When Mercury is glimpsed without a telescope, it is bound to be near the horizon, so that it will be shining through a deep layer of the Earth's atmosphere, and the image will be unsteady. The best method is to find the planet as early as possible, while it is still fairly high up; for sweeping, it is advisable to use either binoculars or else a low magnification on a small telescope (I have found that a power of 25 on a 3-inch refractor does very well). A drawing can then be made while the sky is still bright.

Telescopes fitted with equatorial mountings and clock drives allow a faint object to be found without any tiresome sweeping. This saves a great deal of time, though Mercury is never easy to locate except with a large instrument. Yearly star almanacs tell where and when it is to be seen, and more detailed tables are given in *The Handbook of the British Astronomical Association*.

It is also possible to sweep for Mercury in broad daylight, but it is never wise to range about with a telescope until the Sun has set. Moreover, Mercury and the Sun will not be far apart, and there is always the chance that the Sun will enter the field of view during sweeping, with disastrous results.

For examining the phase and for drawing any visible surface markings, the magnification used should be as high as possible, but the slightest unsteadiness or blurring will be fatal, so that one has to strike a happy mean. All things considered, Mercury is more difficult to study than any other planet, and it is hopeless to expect to see anything spectacular.

Occasionally, Mercury passes in transit across the face of the Sun. When this happens, it can be seen as a well-defined blank disk, quite unlike a sunspot. Transits are of no real importance, but they are interesting, and projection with a 3-inch refractor is quite adequate. The next transits will be those of May 9, 1970 and November 10, 1973.

Mercury is so small and remote that even if it did not remain obstinately close to the Sun, we could hardly hope to find out much about it. Not so with Venus, which is practically the same size as the Earth, and is the nearest body in the sky apart from the Moon. The problems here are quite different, but they are equally puzzling. Though Venus is so close on the astronomical scale, it remains a planet of mystery.

Fig. 35. Apparent size of Venus at various phases.

Venus is a splendid object to the naked eye, and can even cast a shadow at times, but telescopically it is a grave disappointment. When at its most brilliant, it shows as a crescent, since by the time of "dichotomy" (half phase) it has already drawn away from us, so that its apparent diameter is much less (Fig. 35). Altogether, Venus is a most infuriating object.

Moreover, the disk appears virtually blank, even with powerful telescopes. Vague, dusky shadings may be seen often enough, but they are not permanent, and are so diffuse that they are hard to define. In fact, we are not looking at the true surface of Venus at all; we are seeing only the upper layers of an obscuring atmosphere.

Up to the end of 1962 our information about Venus as a world was remarkably slight. Large quantities of carbon dioxide were detected in the upper atmosphere of the planet, and there were reports of water vapour as well as a certain amount of free oxygen, but even a problem so fundamental as the length of the axial rotation period remained unsolved. Various estimates were given, ranging from $22\frac{1}{4}$ hours up to as much as $224\frac{3}{4}$ days; in the latter case the rotation would, of course, be "captured" in the same way as the Moon's with respect to the Earth, so that Venus would keep the same hemisphere turned toward the Sun. Opinions as to the nature of the surface fluctuated wildly between a swampy, tropical hothouse, a planet completely covered with water, and an arid dust-desert without a scrap of moisture anywhere.

All that could really be said was that Venus was likely to be hot, and to be unsuitable for terrestrial-type life except possibly in the form of primitive marine organisms. The only real way to find out more was to dispatch a rocket, and this was what the Americans actually did in 1962. Mariner I was a failure, but Mariner II proved to be a brilliant success.

Mariner was not designed to land upon Venus. The original idea was to send it within 10,000 miles of the planet, keeping in radio touch with it, so that various measures could be made—particularly with regard to the surface temperature. Slight and understandable errors meant that the minimum distance from Venus was more than twice as great as had been hoped, but on December 14, 1962, Mariner passed only about 21,000 miles from the planet. Signals from it were clearly received, and when they were analysed, in early 1963, they certainly provided a sensation. Almost all the findings were unexpected.

First, Venus seemed to have no magnetic field. Secondly, the surface temperature was measured at around +800 degrees Fahrenheit, which at once disposed of the "oceanic" theory; under such conditions water could not exist in the liquid form. And thirdly, the rotation period appeared to be even longer than the sidereal period of $224\frac{3}{4}$ days.

Some authorities still have doubts about the Mariner findings, but the general view at present is that the results are at least of the right order. The latest radar work indicates that Venus has a rotation period of 247 days, and spins in a retrograde or "wrong-way" direction.

Visual studies of the upper clouds are still of value, and here the amateur can make himself extremely useful. There is no point in observing Venus when it is shining brilliantly down from a dark sky; the brilliant disk will be dazzling, and the image is likely to be violently unsteady. I have found that the best seeing is obtained when the planet can just be detected with the naked eye, shortly after sunset or shortly before sunrise, but observations made in broad daylight are almost as good.

Venus is so bright that it can usually be found without much difficulty even when the Sun is above the horizon. In general it will not stand a high magnification, but I have often used 150 on a 3-inch refractor and 250 on a 6-inch reflector.

One obvious line of research is to try to determine the positions of the dusky shadings, and to follow any drifts. Unfortunately the shadings themselves are so impermanent that they can seldom be identified from one night to the next, and so far all observers have had to admit defeat. Yet there are possibilities for the future; sometimes a more persistent shading makes its appearance, and there is always a chance of finding out something definite. In 1924, for instance, the dark patches were much more prominent than is usually the case.

There are also occasional bright areas, and it very often happens that whitish caps are seen in the north and south. Both Earth and Mars possess polar caps, but in the case of Venus the appearance is certainly not due to ice or snow, so that "clouds" of some kind must be responsible. Incidentally, it is unwise to jump to the conclusion that the whitish patches mark the poles of rotation, since we still know nothing about the tilt of Venus' axis. Many estimations have been made, but none can be regarded as trustworthy.

It is also worth while examining the curve of the terminator. Since the orbit of Venus is known with high accuracy, it should be easy to predict the exact moment of "dichotomy", or exact half phase, but oddly enough the predictions are usually wrong by several days. When Venus is an evening star, observed dichotomy is always early; with morning elongations, dichotomy is late. The first astronomer to notice this curious disagreement was Schröter, and I have called it "Schröter's Effect", a term which seemed to have been generally accepted.

There is no chance of Venus being out of position, and the atmosphere of the planet must be responsible in some way, but the exact cause is still uncertain. Timing the actual moment of dichotomy is therefore valuable. The terminator will appear sensibly straight for several nights in succession, so that a series of observations is necessary. What generally happens, of course, is that clouds intervene at a critical stage, and cause one to miss a vital evening's or morning's observation. I have been trying to make accurate determinations ever since 1933, but I have yet to have the pleasure of more than four consecutive clear nights at the right moment. Mercury, incidentally, does not seem to show the Schröter effect, presumably because of its virtual lack of atmosphere; but more measures are needed, and are within the range of the well-equipped amateur.

The terminator itself shows occasional slight projections and indentations. Schröter thought that these were due to differences in level, and believed that he had charted a mountain 27 miles high (!). That irregularities exist is undeniable, and they should always be looked for; some are due to contrast effects, but others may be genuine, though it is most improbable that they are due to anything more substantial than "clouds".

Last, but by no means least, there is the Ashen Light. When Venus is a crescent, the night area can often be seen shining faintly, so that the whole disk can be traced. The same appearance can be seen with the crescent Moon, but the cause is different. With the Moon, the glow is due to reflected earthlight, but the Earth is certainly unable to illuminate Venus, and Venus itself has no moon. Weird theories have been advanced to explain the Ashen Light—in 1840, Gruithuisen suggested that it might be due to general festival illuminations lit by the inhabitants of Venus to celebrate the crowning of a new ruler*—but some authorities dismiss it as a pure contrast effect.

One interesting theory is that it is caused by brilliant auroræ in the upper atmosphere of Venus. Since Venus is closer to the Sun than we are, there is no reason to doubt that auroræ exist, and the explanation is not unreasonable. If we could show that the Light is at its brightest during periods of solar activity, when terrestrial auroræ are frequent, we might be able to clear up the problem. I have made an attempt to analyse the available observations of the Light, but unfortunately the observations themselves are too scattered to be of much use. It must however be added that the "aurora" theory has been weakened by the Mariner II revelation that Venus has no appreciable magnetic field.

All the markings on Venus are so indefinite that they are

* Sometimes the Light may be seen several times per week, so on Gruithuisen's theory the Government of Venus would seem to be somewhat unstable!

hard to record, and there is the added complication that the great brilliancy of the disk tends to produce regular, streaky patterns that do not actually exist at all. The nebulous aspect should be drawn as faithfully as possible, and if the depth or sharpness of a shading is exaggerated—as is sometimes necessary —the observer must always be careful to write an explanatory note beside his sketch. A scale of 2 inches to the planet's diameter is convenient.

Clearly there is much that we do not know about Venus, and the more we probe the more baffled we feel. Even transits are irritatingly rare, and the next will not occur until A.D. 2004. It is a relief to find that Mars, the fourth planet in order of distance from the Sun, does at least present surface markings that we can interpret with some degree of success.

Mars can approach us to a distance of 35 million miles. It is therefore always at least 150 times as remote as the Moon, and it is much smaller than the Earth, with a diameter of only 4,200 miles. Fortunately, it is better placed than either Mercury or Venus. Since it lies beyond the Earth's orbit, it can never appear as a half or a crescent, while at its most gibbous it looks the shape of the Moon two or three days from Full.

The main trouble about observing Mars is that it comes to opposition only at intervals of nearly two years, as explained in Chapter 4. We can see fine details only for a few weeks to either side of opposition,* so that the opportunities for useful work are brief; the observer has to make the most of the limited time at his disposal. Nor are all oppositions equally favourable, because Mars has an orbit that is more elliptical than ours. When opposition occurs with Mars near perihelion, the distance is relatively small; at an aphelic opposition, such as that of 1963, the distance may never be less than 60 million miles. In Fig. 36, the orbits of the two planets are shown, with the opposition positions for 1948 to 1963.

The 1956 opposition was the best of recent years, though from Britain Mars was too low in the sky to be well observed. The opposition of 1958 was almost as close, but that of 1961 was less favourable, while those of 1963 and 1965 were even

* It is often thought that Mars can be well seen only on the actual date of opposition, whereas in fact there is no obvious difference in observing conditions for an appreciable time to either side of the opposition date.

worse. 1967 showed an improvement, and the oppositions of 1969 and 1971 will be close.

Mars has a "year" of 687 Earth days, and the tilt of the axis is much the same as ours, so that the seasonal cycle is similar. Since it has an average distance of about 141 million miles from the Sun, compared with the 93 million of the Earth, we must expect it to be cool; but it is certainly not a frozen world. The maximum summer temperature on the equator may attain 70 degrees Fahrenheit, and though the nights are bitterly cold they are not un-endurable. The axial rotation period is 24 hours 37 minutes, and the surface gravity only 4/10 of ours. If an Earthman stood upon Mars, he would be able to jump more than ten feet above the ground.

The beginner is apt to be disappointed with his first tele-scopic view of Mars. Whereas he may ex-pect to see a vast globe streaked with canals and blotched with

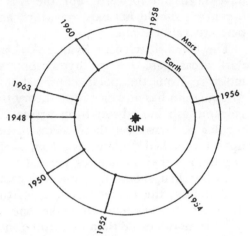

Fig. 36. Oppositions of Mars, 1948-1963. It will be seen that the 1956 opposition was very favourable, as it occurred with Mars near perihelion, whereas those of 1948 and 1963 are most unfavourable.

obvious vegetation areas, he generally sees nothing except a minute reddish disk crowned in the north or south with a whitish cap. It is only when he has become thoroughly used to planetary work that he can make out definite detail. The mark-ings on Mars are much less spectacular than the belts of Jupiter, the rings of Saturn or even the phases of Venus.

The white caps are made up of snowy or frosty deposit; and unlike those of Venus, they are surface features marking the poles. They can be seen with any small telescope when Mars is well placed, and their behaviour tells us a great deal about conditions on our neighbour world. During the winter of either hemisphere, the cap is large and conspicuous, but with the arrival of warmer weather in the spring it begins to shrink. This shrinking goes on well into the summer, and the southern cap has been known to vanish completely for a time.

The shrinking must release a certain amount of water or water vapour, and this has a marked effect upon the darkish patches nearer the equator. They seem to harden and sharpen, as though being revived by the moisture, and it seems therefore that they are due to plant life. There is no proof; other theories have been put forward, but there is no reason why living organisms should not exist on Mars, and the explanation is a perfectly rational one.

The Swedish scientist Svante Arrhenius supposed that the dark areas were due to "hygroscopic salts", which absorbed moisture from the polar caps, while more recently D. B. McLaughlin has advanced the theory that the areas are due to volcanic ash hurled out from active vents and laid down in regular patterns. Yet there seems to be no need to fall back upon far-fetched explanations such as this. The areas look like vegetation; they behave as vegetation might be expected to do; why, then, should they not *be* vegetation?

Certainly, the rocket results from the U.S. Mariner IV—described below—tend to make one slightly less confident; Mars is more hostile than used to be thought. Yet it is difficult to find an alternative to the vegetation theory, and it remains a probability even though it is certainly not conclusively proved.

However, we cannot expect to find trees or bushes, or even flowers. The thin atmosphere of Mars, with a ground pressure no greater than that encountered 18 miles above sea-level on the Earth, is not inviting, particularly as it contains almost no free oxygen. The plants must be as lowly as our lichens and mosses. It is dangerous to draw too close a comparison with our own vegetation, since conditions on Mars are so different from anything in our experience, but we can be fairly sure that higher forms of plant life are absent.

The seasonal cycle is regular. As a cap shrinks, and moisture is released, a "wave of darkening" seems to sweep gradually from pole towards equator. There are also non-seasonal variations in some areas such as the Solis Lacus (see the map given on page 214), due possibly to the temporary spread of plants on to the neighbouring barren areas. Some regions are subject to major alterations in form, but in general the boundaries of the dark patches remain constant from one century to another. The old observers regarded them as seas, but it is now known that there are no oceans anywhere on Mars, and that the whole planet is very short of water.

The ochre areas, which give Mars its ruddy hue and have led to its being named after the God of War, are often called "deserts". This is a reasonable description, but only in a broad sense; the popular idea of a Martian desert as a tremendous Sahara, punctuated with fertile oases and with camels strolling casually among the date-palms, is distinctly wide of the mark. There is no running water on Mars, so far as we know, and so there can be no sand. The deserts appear to be made up of dust covered with some reddish mineral.

Clouds can be seen, but here again we must use the word in its broadest sense. Martian clouds are made not of water vapour, but of dust; rainfall is certainly unknown. The less prominent, fleecier clouds may be composed of high-altitude ice crystals, and if so they will have much in common with the cirrus clouds of our skies, which so often herald the approach of bad weather.

The first good map of Mars was drawn up in 1877 by an Italian astronomer, G. V. Schiaparelli. In addition to charting the dark areas, and giving them the names still in use to-day, he drew numbers of narrow, artificial-looking lines crossing the deserts, and christened them "canali" or channels. Schiaparelli was an eminent observer, and did brilliant work in many astronomical fields, but in the popular mind this one discovery is enough to link his name chiefly with Mars.

Who has not heard of the Martian Canals? Percival Lowell, who built an observatory in Arizona specially to study them, believed them to be artificial waterways built by intelligent engineers to carry water from the icy poles to the arid tropics, so that in his opinion Mars was the home of a highly civilized race. It was natural that his opinions should arouse great interest not only among scientists, but among the general public as well, so that Mars became a topic of every-day conversation.

Unfortunately, there are numerous grave objections to any such idea. To begin with, the polar caps are very thin. Their depth cannot be more than an inch or two, so that they would be unable to provide enough water to fill even one major canal, to say nothing of the 800 charted by Lowell. Moreover, modern analysis of the atmosphere has shown that there is not enough oxygen to support animals or men, while alien life-forms of the kind known to fiction-writers as B.E.M.s (Bug-Eyed Monsters) belong only between the pages of a novel. And lastly, the straight, narrow canals drawn by Lowell seem not to exist in such a form. Under good conditions, observations with large telescopes show that a canal is made up of irregular "fine structure", so that the artificial appearance recorded by Lowell and others is due to the tendency of the human eye to join up disconnected details into hard lines. Since the canals appear to share in the seasonal cycle of the dark areas, it seems possible that they too are due to lowly vegetation. Other astronomers doubt whether the canals exist at all. I can only comment that during the close approaches of 1956 and 1958 I made numerous drawings of Mars with large telescopes without seeing a trace of any sharp, narrow lines. This was also my experience in 1961 and 1963. Certainly, too, no Lowell-type canals were shown on the 1965 rocket probe photographs.

Bad drawings of Mars are regrettably common, even in textbooks. Crude draughtsmanship can be forgiven, but an observer who uses a 3- or even a 6-inch telescope to record dozens of canals is deceiving himself as well as others. It is very easy to "see" what one expects to see, and for this reason it is best to go to the telescope with a completely open mind. Tables given in *The Handbook of the British Astronomical Association* can be used to work out the longitude of the central meridian for any particular time, but such calculations should be made after the observation and not before.

Since drawings of Mars have to be made with comparatively high magnifications, the planet is a difficult object for small telescopes. A 3-inch refractor will show the caps and some of the main dark areas, such as the Syrtis Major and the Mare Acidalium, but for useful work at least a 6-inch is needed, while an equatorial mount and a clock drive should be added if possible.* A scale of 2 inches to the planet's diameter is customary; when the phase is evident, as is always the case except near opposition, the disk should be drawn to the correct shape.

Begin, as always, by looking carefully at Mars for some time until your eye has become thoroughly prepared. Then sketch in the main details, the caps and dark areas, using a moderate power. Change to the highest magnification that will give a sharp, steady image, and fill in the finer detail. As soon as this has been done, note the time, and make a record of it below your sketch. This is important; Mars spins on its axis in 24 hours 37 minutes, so that the drift of the markings across the disk becomes noticeable over even short periods. (Obviously, any particular marking will pass over the central meridian of Mars about half an hour later each night, since the rotation period is half an hour longer than ours.) Finer details can then he added without undue haste. Colours, intensities of the dark areas, and any clouds should always be looked for, as well as features such as a dark border to the polar cap, seen when the cap is shrinking and attributed to temporary moistening of the ground.

Very small telescopes are useless for serious work on Mars, but oddly enough it has been stated that giant instruments also are unsuitable. Sidgwick states† that "if the aperture exceeds about 12 inches, the atmosphere will seldom allow the full aperture to be used". This is a well-worn argument, but it is completely false. It is true that an increased magnification will also increase any tremor due to the air, but under normal conditions a large telescope will always show more than a small one. This has been my own experience with instruments ranging from a 3-inch refractor up to a 33-inch, and it is significant that E. M. Antoniadi, whose work has formed the basis of most modern investigations of Mercury, Venus and Mars, used the Meudon 33-inch for his main research without the slightest temptation to stop down the aperture. However,

* In his excellent book *Observational Astronomy for Amateurs*, J. B. Sidgwick states that an equatorial is "a necessity" and a drive "virtually so". This is certainly incorrect. W. F. Denning, one of the greatest planetary observers of the late nineteenth century, always used a telescope mounted on a simple altazimuth, and was awarded the Gold Medal of the Royal Astronomical Society for his work.

† J. B. Sidgwick, *Observational Astronomy for Amateurs*, Faber & Faber, London, 1955. Page 117.

a reflector of from 8- to 12-inches aperture is enough to allow the amateur to play his part in the observing programme for Mars. Drawings made with smaller apertures are bound to be rather suspect.

Owners of larger instruments may care to look for the two tiny moons, Phobos and Deimos. Both are veritable dwarfs less than a dozen miles in diameter, so that even when Mars is near opposition they are difficult to glimpse. I have seen them both with a 15-inch reflector, and keener-eyed observers should catch sight of them with a 12-inch when conditions are first-class.

A rather stupid mistake on my part may serve to show that it is not wise to reject an observation because it does not "fit in" with what is expected. I was once observing Mars with my $12\frac{1}{2}$-inch reflector, when I recorded a minute starlike point, clearly visible only when Mars itself was hidden by an occulting bar, which I took to be Phobos. I then consulted my tables, and found that Phobos was not in fact anywhere near the position recorded. I therefore dismissed the observation, as either a mistake or else an observation of a faint star. It was only on the following day that I found that the observation itself was perfectly correct; I had made a slip in my calculations.

Phobos is a peculiar little body. It whirls round Mars at a distance of only 3,800 miles above the surface, about as far as from London to Aden, and it completes one revolution in only $7\frac{1}{2}$ hours. So far as Phobos is concerned, the "month" is shorter than the "day", and to a Martian observer Phobos would seem to rise in the west, gallop across the sky—taking only $4\frac{1}{2}$ hours to pass from horizon to horizon—and set in the east. Neither it nor Deimos would be of much use as a source of moonlight, and Deimos would indeed look like a large, dim star.

Though Mars is farther away than Venus, it is within the range of modern unmanned rockets, and the first Mars Probe was launched in 1962 by the Russians. Unfortunately a fault developed in the probe during flight, and contact with it was lost—a familiar Russian weakness; they also lost their two Venus probes of 1965–6, though it is very likely that the second of these actually landed on the planet.

Where the Russians failed, the Americans succeeded. In mid-1965 their probe Mariner IV passed within a few thousand miles of Mars, and sent back the first close-range photographs of the surface. Here, too, the results were startling. No canals were seen (which was only to be expected), but numerous craters showed up, bearing a remarkable similarity to those of the Moon. In fact, Mars proved to be much less like the Earth, and much more like the Moon, than had been anticipated. It was also found that the Martian atmosphere is considerably thinner than the earlier measures had indicated.

Though the Martian craters are large by Earth standards, they are quite beyond the range of any ground-based telescopes, and without rocket research they could never have been discovered. At present it is too early to say more, but there can be little doubt that new probes will be launched in the near future.

Many books have been written about Mars, and there is so much to be said that the outline given here is bound to be sketchy and incomplete. It may however help to indicate lines of work that can be carried out by the amateur, and though Mars may not be an easy object it is certainly a fascinating one. We must admit that our views have changed of late. Before mid-1965, it was thought that Mars was not too unlike the Earth, with an atmosphere made up largely of nitrogen, and surface conditions which were no more than somewhat hostile. Mariner IV has shown that the atmosphere is much thinner than had been expected, and it now seems that carbon dioxide is the main constituent; the atmospheric mantle may be inefficient as a screen against dangerous radiations coming from the Sun, and altogether Mars is much more hostile than had been hoped. Yet our knowledge is still incomplete, and we may yet have many surprises in store for us.

Chapter Ten

THE OUTER PLANETS

As soon as we look at a scale map of the Solar System, it is seen that the division of the planets into two main groups is very pronounced. Between the orbits of Mars and Jupiter there is a wide gulf of over 300 million miles.

Nearly 200 years ago, Johann Bode suggested that there might be a small planet revolving round the Sun at a distance of about 260 million miles. There were sound reasons for believing that he might be right, and towards the end of the century a group of six leading astronomers, headed by Schröter and the Baron von Zach, began a systematic search for the missing body. Oddly enough, they were forestalled. Before the scheme was in working order, Piazzi at Palermo happened upon a starlike object that turned out to be a small world circling the Sun at almost the correct distance. It was named Ceres, in honour of the patron goddess of Sicily.

Ceres is a dwarf world only 427 miles in diameter, so that it must be totally without atmosphere, and is a mere lump of rock devoid of any kind of life or activity. But it seemed too insignificant to be a major member of the Sun's family, and Schröter and his "celestial police" continued with their programme. Between 1801 and 1808 they discovered three more minor planets, and when a fifth was added in 1845 it became clear that the original four were merely the brightest members of a whole shoal. Since 1848 no year has passed without fresh discoveries, and over 2,000 of these minor planets or "asteroids" are now known, while the total number has been estimated as at least 40,000.

Ceres remains the largest known of the swarm, and of the rest only No. 2, Pallas, has a diameter exceeding 250 miles. Some are real midgets less than a mile across, so that there is no definite distinction between a very small asteroid and a very large meteor. Of all the minor planets, only No. 4, Vesta, can be seen with the naked eye when at its brightest. The rest are always invisible without a telescope.

Hunting and photographing asteroids is a pleasant pastime, and it is not difficult. I once spent an evening searching for known asteroids with a 6-inch refractor, and observed fifteen of them in only two hours, though I could not identify them all until I re-observed on the following night.

The procedure is to look up the position of a suitable asteroid, using an almanac or the B.A.A. *Handbook*, and plot it on your star chart. Then go to the telescope, and search until you have found the desired star-field, using the method described in

Fig. 37. Apparent shift of the minor planet Pallas over a period of 24 hours; Patrick Moore, 3-in. refractor.

Chapter 15. As the minor planet will look exactly like a star, it will not be recognizable at first sight, so the only course is to make a map of all the stars in the area. When you look again the following night, the stars will be unchanged, but the minor planet will have betrayed itself by its obvious shift in position. Two drawings of this kind are shown in Fig. 37.

Though most of the minor planets remain in the main zone between Mars and Jupiter, some have unusual paths. The "Trojans" are exceptionally remote, and have the same mean distance as Jupiter, so that they are very faint, whereas the extraordinary asteroid Hidalgo has an eccentric orbit that carries it from inside the path of Mars out almost as far as Saturn. As well as being so elliptical, Hidalgo's orbit is tilted to the ecliptic at the high angle of 43 degrees.

Even stranger are the occasional minor planets which make close approaches to the earth. Eros, the largest of them, has a minimum distance of 15 million miles, and has been most useful in helping astronomers to measure the length of the "astronomical unit", the mean distance between the Earth and the Sun, though admittedly the Eros method has now been superseded. Other "Earth-grazers" can come even closer. The present holder of the record is Hermes, only a mile in diameter, which whirled by us in 1937 at a distance of only about 400,000 miles. This is still twice as far away as the Moon, but when the news was released there were some people who became really alarmed at the idea of a celestial collision, while one national newspaper produced the immortal headline: "World Disaster Missed by Six Hours." Actually, the chances of our being hit by an asteroid of any size are so small that they can be neglected.

Oddest of all the minor planets is Icarus. At aphelion it lies beyond the orbit of Mars, but at perihelion it swings to within 19 million miles of the Sun. It is then closer than Mercury, and its "day" side must be red-hot. In 1968 it will approach the Earth to within a few millions of miles, but there is no danger of a collision, despite some recent sensational reports in the Press!

Beyond the main asteroid zone we come to mighty Jupiter, giant of the Solar System. Though it never approaches us much within a distance of 360 million miles, well over a thousand times as remote as the Moon, Jupiter still shines so brilliantly in our skies that it cannot be mistaken for a star. It is outshone only by Venus and, very occasionally, by Mars.

Jupiter's vast globe could contain 1,300 bodies the size of the Earth, but it is not so massive as might be supposed. If we could put Jupiter in one pan of a pair of scales, we should need only 318 Earths to balance it. This must mean that Jupiter is less dense than the Earth, and the density works out at only 1·3 times that of water.

Jupiter is not a rocky body like the Earth or Mars. When we look at it through a telescope, what we see is not a hard surface, but a cloudy vista with details which change not only from night to night, but from hour to hour. We must not, however, draw any comparison with Venus. Jupiter's "atmosphere", to use the word in a broad sense, is so deep that it merges with the true "body" of the planet.

It used to be thought that Jupiter consisted of a rocky core, overlaid by a 15,000-mile thick layer of ice which was again overlaid by the atmosphere. Recent research has cast doubts upon this theory, and it is possible that the whole globe is made up chiefly of hydrogen, so compressed that near the centre it starts to behave like a metal. Both ideas may be wrong; time will tell.

At any rate, we can carry out analysis of the upper gas, which proves to be an unprepossessing mixture of ammonia, methane and free hydrogen. Both ammonia and methane are poisonous, and when it is remembered that the temperature on Jupiter can never rise above −200 degrees Fahrenheit we can see that any form of life there is out of the question.

In a small telescope, Jupiter appears as a yellowish disk, flattened at the poles and crossed by prominent streaks known as "belts". Increased power shows finer details such as wisps, brightish areas and spots. Though all these are phenomena of the high atmosphere, studies of them can tell us much about Jupiter itself, and amateur work in past years has been of the greatest value. The records of the Jupiter Section of the British Astronomical Association, directed for many years by the Rev. T. E. R. Phillips and now by W. E. Fox, are the most complete in existence.

The belts, due probably to droplets of liquid ammonia, dominate the picture. Usually there is a prominent belt to either side of the equator, while moderate telescopes will reveal others. They vary in prominence, as becomes evident if observations are continued from year to year. In 1962–1964 the general aspect was most unfamiliar, since the two equatorial belts ran together to form a dark band, but by 1966 the appearance was back to normal.

Spots are generally short-lived, and last only for a brief period before disappearing. The chief exception is, of course, the famous Great Red Spot, which beca me very striking in 1878 and can be traced on drawings made as early as 1631. In its prime, the Spot was a brick-coloured elliptical object 22,000 miles long and 7,000 wide (Fig. 38), and was still conspicuous in 1967, though it has disappeared at times. Its origin is still a mystery. It has been attributed to a volcanic eruption, but this idea has now been discarded, and all we do know is that it may be a semi-solid body or group of bodies lying in the upper atmosphere of Jupiter. It is the only feature of the disk which has been known to persist for more than half a century, its nearest rival in this respect being a disturbance in the south tropical zone which lasted from 1901 to 1940.

If Jupiter is watched for a few minutes with a magnification of 150 or more, the surface features will be seen to be drifting slowly from right to left. This is the result of the planet's axial spin, and is more obvious than in the case of Mars, since Jupiter has a much shorter "day". In the tropical zone, between the two equatorial belts, the period is only 9 hours 50½ minutes, while in higher latitudes it is 5 minutes longer. Jupiter does not rotate as one mass; different zones have different rates of rotation, and this is an extra proof that the surface we see is not a solid body.

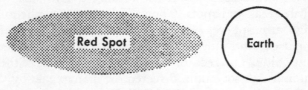

Fig. 38. Size of the Great Red Spot.

Moreover, individual features have individual rotation periods. Between 1901 and 1940 the Red Spot and the South Tropical Disturbance were both to be seen, and the Disturbance moved the more quickly of the two. Periodically it caught up the Spot and "lapped" it, and when the two were close together they seemed to interact. (In 1966 what seemed to be a new South Tropical Disturbance was detected by T. J. C. A. Moseley at Armagh; I saw it shortly afterwards, and we had high hopes of it, but to our disappointment it faded away after a few weeks.) In 1919–20 and in 1931–34, observers of the B.A.A. Jupiter Section even observed "circulating currents" in the south tropical zone, and many other interesting examples could be given.

Jupiter's quick rotation means that one cannot afford to linger when making a disk drawing. The sketch should be completed in less than 10 minutes, as otherwise the drift of the surface features will introduce errors. As in the case of Mars, the main details should be filled in first; the time should then be noted, after which the magnification can be increased and the finer details added.

One minor irritation is that one cannot use a pencil compass to draw the outline of the disk. The polar compression amounts to 6,000 miles (as against 26 miles in the case of the Earth) so that it cannot be neglected, and shaping the outlines free-hand is a tedious process. I have found that the best solution is to obtain a stock of blanks, as shown in Fig. 39. These blanks are not expensive to have printed, and any local firm will make them at low cost.

Fig. 39. This diagram can be traced and a line-block obtained so that a printer can run off a stock of blanks.

Rotation periods of special features are best determined by the method of transits. There is no analogy with the solar transits of Mercury and Venus, and the word is used to denote the time when the feature under study passes across the central meridian of Jupiter.

What is done is to estimate the time of transit to the nearest minute, which is quite adequate. A measuring device is naturally helpful, but visual estimates can be made quite accurate enough for most purposes, and Jupiter rotates so rapidly that it is often possible to time 20 or 30 transits per hour. There is a standard nomenclature, and this is given in Appendix VII, together with an extract from my own notebook that may prove helpful. It is hardly necessary to add that a reliable watch is essential—and make sure that it is set to the correct G.M.T.!

Once the time of transit has been found, the longitude of the feature can be found by means of the tables in the B.A.A. *Handbook*. This is an easy process, and involves nothing more frightening than simple addition.

Transits assumed unexpected importance in 1955, when two American researchers, B. F. Burke and K. L. Franklin, found that Jupiter emits long-wave radiation of the type known scientifically as "radio noise". The discovery was surprising, and the radio astronomers naturally wanted to know whether the emission came from the whole planet, or merely from small active regions of the disk. If the latter, the radio emission should be at its most powerful when the feature concerned is on the central meridian. It is still rather early to come to any definite conclusions, but work is in progress, and more transit observations are urgently needed.

Neither do we know what causes the radio emission. It has been suggested that intense thunderstorms occur on Jupiter, but no thunderstorm could provide as much energy as is actually emitted. Strangely, the position of Io, the innermost large satellite of Jupiter, seems also to be involved. Further research is now being carried out.

For routine work on Jupiter, a power of 150 to 250 on a 6-inch reflector is adequate. Transits can be taken as accurately as with a larger instrument, but there will be fewer observed, since only the major features of the disk will be visible.

As befits the senior planet of the Solar System, Jupiter has a retinue of twelve moons or satellites. Four of them are bright enough to be seen with any telescope, and there are records of their having been seen with no optical aid at all, but the remaining eight are too dim to be glimpsed with any amateur-owned equipment.

Astronomers have been quick to christen even the most unimportant asteroids, but oddly enough the satellites of Jupiter are still officially anonymous. The names of Io, Europa, Ganymede and Callisto for the four chief moons have at last come into general use, while No. 5 is known to Continental astronomers as Amalthea, but the rest are still mere numbers. This leads to confusion; Moon No. 12, for instance, is closer to Jupiter than No. 9.

This seems rather a pity. Suitable names could be found easily enough, and Marsden in the *B.A.A. Journal* has proposed Hestia, Hera, Poseidon, Hades, Demeter, Pan and Adrastea for Numbers 6 to 12. These fit well into the mythological mood, and I propose to use them in the present book, with the full knowledge that I am courting official disapproval.

Io and Europa are about the size of our Moon, while Ganymede and Callisto are larger, and actually of greater diameter (though lesser mass) than Mercury. No atmospheres round the satellites have been definitely confirmed, and surface details can be seen only with great telescopes, but the movements are fascinating to watch.

Since all four revolve approximately in the plane of Jupiter's equator, they appear to keep in almost a straight line, but it often happens that one or more of them is missing. A satellite may pass in front of Jupiter, appearing in transit;* it may pass behind, and be occulted; it may pass into Jupiter's shadow, suffering eclipse. The transits are particularly striking. In Plate VIII(c), a typical view, the dark disk of Ganymede is seen against the Jovian clouds. Accurate timing of these phenomena is valuable. All these transits, eclipses and occultations are predicted for each year in the B.A.A. *Handbook*, and in many almanacs.

The remaining eight satellites are among the faintest observable objects in the Solar System. Amalthea, which lies closer to Jupiter than any other member of the retinue, has been recorded with an 18-inch under the best possible conditions, but Adrastea has never been "seen" visually, though it has left its image on photographic plates. The diameters range from 150 miles (Amalthea) down to only about 14 miles (Adrastea), so that they are inferior to many of the asteroids. Their orbits are strange; the outer four are so distant from Jupiter that they take over a year and a half to complete one revolution, while to make matters even more complicated

---

* This is yet another use of the word "transit". A satellite transit has nothing to do with the apparent passage across Jupiter's central meridian.

Adrastea, Pan, Poseidon and Hades go round the wrong way, east to west instead of from west to east. These four are so far out that even Jupiter's mighty pull is barely sufficient to control them, and consequently their orbits are not even approximately circular. Poseidon, discovered in 1908, was actually "lost" for some time after 1941, and was not found again until 1955. Possibly these moonlets are not true satellites at all, but merely minor planets that have been captured by Jupiter and forced to give up their independent status.

Far beyond Jupiter, at an average distance of 886 million miles from the Sun and a minimum of 741 million from the Earth, lies Saturn, second of the giant planets. In itself Saturn is less important than Jupiter; it is smaller, with an equatorial diameter of 75,100 miles and a mass of 95 times that of the Earth, and it is made up in much the same way. It is even colder than Jupiter, since the temperature never rises above − 240 degrees Fahrenheit, and it too must be utterly lifeless.

Saturn shows belts and spots, but surface features are much less conspicuous than those of Jupiter, and well-marked spots are very rare. The last really spectacular outbreak took place in 1933, when W. T. Hay (Will Hay), a famous comedy actor who was also a skilled amateur astronomer, discovered a short-lived white spot near the equator. I detected a fainter white spot in 1962, but it never became prominent, and soon faded away. Features of this kind can be used for transit observations, as in the case of Jupiter, but they are so unusual that our knowledge of Saturn's rotation period is far from complete. The value for the equatorial zone seems to be 10 hours 14 minutes.

Saturn is a quieter world than its giant brother, but the various zones seem to show changes in brightness, so that intensity observations are of value. These can be made by eye estimates, on a scale of 0 (brilliant white) to 10 (black shadow). The work needs a telescope of at least 8 inches aperture, but fortunately Saturn is a convenient object inasmuch as it will usually stand a comparatively high magnification.

But the glory of Saturn lies in its ring system. Huygens, the leading telescopic worker of the seventeenth century, described it as "a flat ring, which is inclined to the ecliptic and which nowhere touches the body of the planet", but actually there are three rings, two bright and one dusky (Fig. 40). The whole system has a diameter of almost 170,000 miles.

Saturn is a massive planet, and it has a strong gravitational pull. Were the rings liquid or solid, they would soon be broken up and destroyed, so that they must be made up of individual particles whirling round Saturn like miniature moons. It is possible that they are the shattered remnants of a former satellite that wandered too close to its master.

A 3-inch telescope will show the rings, but in a 6-inch the sight is glorious indeed, and Saturn is without doubt the most superb object in the heavens. It is unique in its glory, and it is a sight never to be forgotten.

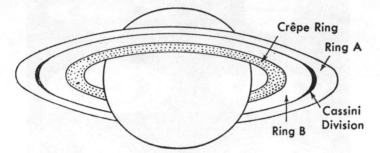

Fig. 40. Diagram of Saturn's ring system.

Details can be seen in the ring-system. The bright rings, A and B, are separated by a dark area known as Cassini's Division, in honour of its discoverer. The Division is a true gap, and is due to the disturbing influences of Saturn's satellites. There is a second gap in Ring A (Encke's Division) which can be seen under good conditions with an 8-inch reflector, and other divisions have been reported, though they have not been fully confirmed and their existence is doubted by many observers.*

Though the rings cover so vast an area, they are strangely thin, far thinner relatively than a sheet of tissue paper. They cannot have a thickness of more than 50 miles, and 10 miles is probably nearer the truth, so that when they are placed edge-on

* I have looked for these divisions with telescopes ranging from 10 to 33 inches aperture, but have seen only Cassini's and (occasionally) Encke's. Neither have I been able to find a fourth "dusky" ring lying outside Ring A, reported on various occasions since 1909. My observations of Saturn with large telescopes are however too few in number to be of much use.

to the Earth they almost disappear. The drawings in Fig. 41 show the alterations in appearance from year to year. The rings were edge-on in 1950, and again in 1966, while they were fully displayed in 1958.

It is not easy to keep track of the rings when the system is exactly edge-on. Small telescopes will show no trace of the rings for a period of several weeks, but during 1966 T. J. C. A. Moseley, P. G. Corvan and myself, using the 10-inch refractor at Armagh Observatory, found that the rings could be seen as

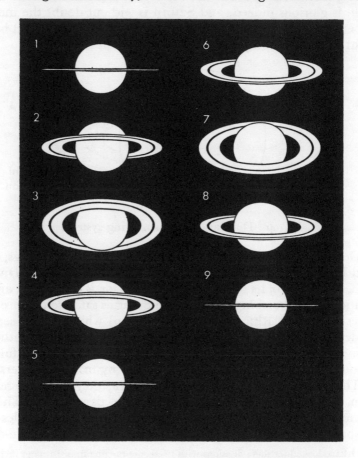

Fig. 41. Aspects of Saturn's Rings. One full cycle is shown; the rings are closed in positions 1, 5 and 9; the southern face of the ring is shown in 2–4, the northern face in 6–8.

a thin and excessively faint line. I doubt whether any smaller telescope would have shown them at all between late October and the end of the year.

Saturn is an awkward object to draw, but there is no "short cut", as in the case of Jupiter. Stencils can be made to allow for the polar flattening of the disk, but the rings have to be sketched freehand. Unfortunately it is not possible to prepare one standard drawing and use it as an outline for weeks on end, as the presentation of the rings alters perceptibly even over short periods.

Points to note are the intensities of the various rings (B is always brighter than A), the shadow effects of rings on disk and disk on rings, and the visibility of any of the Divisions. Occasionally Saturn occults a star, and these occultations are important, since even the bright rings are semi-transparent and the dimming of the star is a key to the composition of the rings. Ring C, the Crêpe or Dusky Ring, has been suspected of variations in brightness.

Saturn has ten satellites. Of these the largest is Titan, 3,500 miles in diameter and hence the largest satellite in the Solar System. It has an atmosphere, made up principally of methane, and can be seen with any 2-inch telescope. Of the remaining satellites, Iapetus, Rhea, Dione and Tethys can be caught with a 4-inch; Enceladus, perhaps Mimas and Hyperion, with a 12-inch reflector, while the remaining satellites, Phœbe and Janus, are much more difficult. Iapetus is interesting because it is at its brightest when to the west of Saturn, and must have a patchy surface of unequal reflecting power.

Saturn was the outermost of the planets known to the ancients. It seemed to move slowly across the starry background, and to them it was dull, heavy and baleful. They certainly gave no thought to the possibility of a more remote planet, and William Herschel's discovery of Uranus in 1781 came as a major surprise to the scientific world.

Herschel was busy "exploring the heavens", to use his own words, and he was not making any particular search for a new planet, so that even when he came across an object which was certainly not a star he did not immediately realize what it was. He mistook it for a comet, but a few weeks' observation proved its true nature. It is interesting to recall that the telescope used by Herschel was a reflector of his own construction.

Uranus is a giant world with a diameter of 29,300 miles, rather less than half that of Saturn. In spite of its great distance, never less than 1,600 million miles from the Earth, it can just be seen without a telescope, and a small instrument will reveal its dim, greenish disk. Faint belts can sometimes be seen with apertures of 10 inches and over, but little else can be made out.

Uranus is a celestial oddity. Whereas most of the planets have their axes of rotation inclined to the perpendicular to the planes of their orbits by 20 or 30 degrees (Fig. 42), Uranus has an inclination of more than a right angle. Consequently the "seasons" there must be most peculiar, particularly as the "year" is 84 times as long as ours. First much of the northern

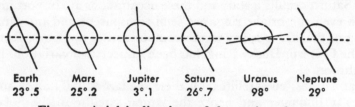

Earth	Mars	Jupiter	Saturn	Uranus	Neptune
23°.5	25°.2	3°.1	26°.7	98°	29°

Fig. 42. Axial inclinations of the major planets.

hemisphere, then much of the southern is plunged into darkness for 21 years at a time, with a corresponding period of daylight in the opposite hemisphere. Sometimes we look straight at the pole, as in 1945, while at others the equator is presented.* In itself, the planet appears to be a smaller edition of Jupiter or Saturn, with the same unpleasant constitution and an even lower temperature.

Few amateurs will possess telescopes large enough for studying the surface of Uranus, but it is interesting to estimate the planet's brightness, since there seem to be irregular variations which may be linked with disturbances on the disk. The method of estimation is to compare Uranus with a near-by star of known brilliancy, just as is done in the case of a variable star (see Chapter 15).

A low power, 50 to 70 on a 3-inch refractor, is best for this work. With higher magnifications, Uranus appears as a definite disk, and is difficult to compare with a star. In 1955, when

* According to an old fairy story, the Earth's axis used to be upright, but the hideous crimes of mankind caused it to tilt to its present angle of 23½°. In the case of Uranus, the tilt is 98°, so that I hate to think what must have happened there!

Uranus and Jupiter lay close together in the sky, I tried to compare Uranus with Ganymede and Callisto, but the planet was so much larger and dimmer than the satellites that I was unable to get any reliable results. Needless to say, the two worlds were not really any closer than usual, but merely happened to lie in about the same line of sight. Uranus is more than three times as remote as Jupiter.

Uranus has five satellites. Of these, Titania and Oberon should be visible with an 8- or 9-inch telescope; Ariel and Umbriel require at least 18 inches, and Miranda is beyond any but the world's largest instruments. Titania, the most easily detected member of the family, is about 1,500 miles across, so that it is appreciably smaller than own Moon.

Neptune, last of the giants, is the true twin of Uranus. It is slightly smaller, slightly more massive, and much more distant, since even at its closest point to the Earth it is still 2,675 million miles away. It can be seen with any small telescope, but with anything less than 4 inches of aperture it looks very like a star. Larger instruments show a bluish disk, practically devoid of detail.

The story of Neptune's discovery is one of the most interesting in astronomical history, since the planet was tracked down before it was actually seen. Between 1781 and 1830, mathematicians found that the new planet Uranus was wandering from its predicted path; it was not moving as it should do, and an amateur, the Rev. T. J. Hussey, suggested that the cause of the trouble might be an unknown body, pulling on Uranus and dragging it slightly away from its expected position. Two investigators, John Couch Adams in England and Urbain Le Verrier in France, set themselves to work out the position of the disturbing body. It was a true detective problem; they knew the victim, and they had to find the culprit.

Adams finished first, and sent his calculations to the then Astronomer Royal, Sir George Airy. Unfortunately Airy took no immediate action, and by the time he did give orders for a search it was too late; Le Verrier's results enabled two German astronomers, Galle and D'Arrest, to identify the new world very close to the position that had been indicated.

Neptune does not share Uranus' great axial tilt, and although it is satisfying to find the remote, frigid giant, there is little scope for the amateur. However, a 6-inch telescope should show the major satellite, Triton, which is brighter than any of the moons of Uranus, and was discovered shortly after Neptune itself had been recognized. The second satellite, Nereid, is excessively faint.

With the discovery of Neptune, the Solar System was once more regarded as complete. Yet the movements of the outer planets were still not in full agreement with calculation; and Percival Lowell, famed for his studies of the Martian canals, undertook to work out the position of a ninth planet.

The problem was much the same as that which had confronted Adams and Le Verrier, but was even more difficult, and Lowell had no success. He died in 1916, but the search was continued at his observatory, and fourteen years later Clyde Tombaugh detected a dim, starlike object which proved to be the missing planet. It was christened Pluto, and the name is apt; Pluto was King of Darkness, and the world named after him must be a dismal, twilight place, with the Sun looking like nothing more than a tiny though intensely brilliant disk.

Pluto has set astronomers problem after problem. The most annoying thing about it is its size. It is much smaller than Lowell had anticipated, and it seems indeed to be no larger than the Earth, so that it is a solid body and not a gaseous globe. It cannot have a strong gravitational pull, and unless something is badly wrong with the measurements it can have no detectable effects upon the movements of Uranus or Neptune —yet it was by these very effects that Pluto was tracked down!

It is hard to believe that Lowell's accuracy was due to sheer luck, particularly as independent work by another American mathematician, W. H. Pickering, gave a similar result. It has been suggested that Pluto is really larger than the measures indicate, but at present the puzzle remains unsolved.

We know little about Pluto itself. Researches carried out in 1956 yield a rotation period of 6 days 9 hours, but the planet is so small and so far away that no ordinary telescope will show its disk.

The orbit is strange, and quite unlike that of any other major planet. The sidereal period is 248 years, and the distance from the Sun varies from 2,766 million miles at perihelion to as much as 4,566 million at aphelion. At its closest to the Sun, Pluto is actually closer-in than Neptune, but the orbit is appreciably tilted, so that there is no fear of the two planets meeting in collision—though there is a possibility that Pluto is a former satellite of Neptune that has broken loose, and is now masquerading as a planet in its own right.

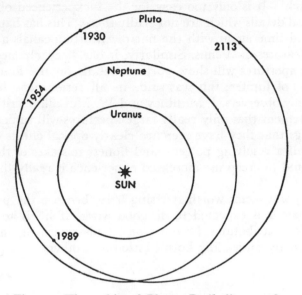

Fig. 43. The orbit of Pluto. Perihelion—1989; aphelion—2113.

Pluto is drawing in to perihelion, and it has brightened up considerably since its discovery. It will go on increasing in brilliancy until it passes perihelion in 1989. A 12-inch telescope will now show it, and it can be identified in the same way as an asteroid, though with more difficulty on account of its slower motion. Fig. 43 shows its orbit. It is worth while making a search for Pluto just for the satisfaction of seeing it.

Is there another planet beyond Pluto? There may well be, but if so it will be so dim that we may never find it. So far as we can tell at present, Pluto marks the frontier of the Sun's inner kingdom.

Enough has been said to show that any amateur with energy and patience can do valuable work in the field of planetary observation. He may not have a large telescope; he may not possess a scientific degree, but at least he can make himself useful if he wants to. And after a lifetime's work, he will realize that there is still much that he has left undone.

There is one danger facing him, however: that of "seeing too much". It is only too easy for the inexperienced observer to record details which are not really there. This has happened time and time again with the narrow Martian canals and the linear features on Venus. Similarly, it has been claimed that modest apertures will show surface details on the four large satellites of Jupiter; this was said, in all seriousness, by two Allegheny observers, J. Mulliney and W. McCall, in criticizing my statement that only really large apertures will suffice. The markings that they have seen are clearly optical effects due to too small a resolving power—and honest mistakes of this sort are bound to creep in; increased experience is really the only cure.

The point seems worth stressing here, because the present-day amateur is expected to do good work. If he is to make valuable contributions, he must be on his guard against unintentionally misleading both himself and others.

Chapter Eleven

COMETS AND METEORS

A BRILLIANT COMET, with a tail that stretches half-way across the sky, is one of Nature's greatest spectacles. Small wonder that it caused fear and panic in ancient times, when comets were believed to be heralds of disaster. Shakespeare wrote in *Julius Cæsar*:

"When beggars die, there are no comets seen:
The heavens themselves blaze forth the death of princes,"

and even to-day the feeling is not entirely dead. Yet there is not the slightest foundation for it, since comets are the flimsiest and most harmless members of the whole Solar System.

Broadly speaking, a comet is made up of small pieces of matter, ranging in size from sand-grains to blocks bigger than houses, enveloped in thin gas. A comet is not therefore a hard, solid body like a planet, and even the largest comet has a mass smaller than that of a minor satellite such as Phœbe.

A few people still confuse comets with meteors, or shooting-stars. There is of course a link between the two, as will be shown below, but there is no excuse for any misunderstanding. Whereas a shooting-star is a piece of matter that dashes into the air, perishing in a streak of radiance after a few seconds, a comet may remain visible for months, moving so slowly against the starry background that its motion cannot be detected except over a period of some hours.

The popular idea of a comet is of a vast fuzzy mass with a magnificent tail streaming out of it. Great comets do in fact look like this, and are made up of three main parts known as the nucleus or central condensation, the coma and the tail, but smaller specimens are much less imposing. I remember showing a telescopic comet to a friend of mine who knew little about astronomy and cared less. His comment was that the comet looked "like a small lump of cotton-wool", and there was some truth in the description.

The coma or head of a comet looks like a filmy mass, and is

made up chiefly of tenuous gas enveloping scattered pieces of meteoric matter. Near the middle of the coma there may be a central condensation, so sharply defined that it looks like a star, and in which the solid particles are more numerous and more closely packed. If there is a tail, it streams away from the coma, merging into it so perfectly that it is usually impossible to tell where the one begins and the other leaves off. The gas composing the tail is so thin that its density is negligible according to our normal standards, but here too there is plenty of dust.*

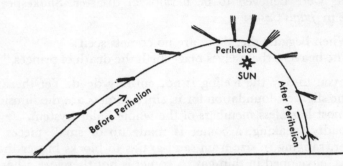

Fig. 44. Direction of a comet's tail with respect to the Sun.

The nucleus or central condensation seems to be the most important part of any comet, and it is assumed that the coma and tail are formed by gases given off by the matter in the nucleus when heated by the Sun. It is significant that most comets develop tails only when near perihelion, and lose them again when they have receded some way after perihelion passage; but there are of course exceptions to the general rule, and we have to admit that the mechanism of tail formation is still not properly understood.

One curious fact is that the tail always points away from the Sun. When the comet is racing towards perihelion it travels head-first in the conventional way, but after passing perihelion it moves tail-first (Fig. 44). The tail must whirl round at a tremendous speed during the passing of perihelion, and in some cases the old tail disappears, to be replaced by a new one on the far side of the comet.

* "Dust" must not be taken to mean the sort of dust that one finds on the mantelpiece in a disused room. The particles in a comet may be mainly ices, as was suggested by F. L. Whipple in 1950.

This interesting behaviour was formerly explained as being due to the fact that light exerts a pressure; it was thought that with the small particles making up a comet's tail, light-pressure was able to drive the material outward. It has now been found that this explanation is inadequate, and that the phenomenon is better accounted for by introducing magnetic effects together with particles sent out by the Sun, though further researches are in progress. At any rate, a comet must be subject to a steady wastage of material, and on the cosmical time-scale it is a short-lived body.

Most of the periodical comets of short period are too faint to be seen without a telescope, even when at their brightest, and when far from perihelion they cannot be seen at all. We speak of the "return" of a comet when it comes back to the regions in which it can be observed. Encke's Comet, for instance, has a period of 3.3 years; it has now been observed at over forty returns, the latest being that of 1967. It is so named because the German mathematician Encke was the first to realize that it revolves round the Sun, and that in consequence its returns can be predicted. Its orbit is shown in Fig. 45.

About 30 known short-period comets, including Encke's, have their aphelion points at or near the orbit of Jupiter. They form a sort of family, and clearly the Giant Planet is concerned in some way. It is not suggested that comets are formed from Jupiter or by Jupiter, but the powerful gravitational pull exerts some control on their movements. By the beginning of 1967 there were about 100 comets known to have periods of less than 200 years, but most of them are extremely faint, so that large telescopes are needed to show them.

The chief exception is, of course, Halley's Comet, which comes back every 76 years. It is the only comet of short or moderate period which can be called "great", and it is a majestic spectacle for a few months at each apparition. It is named after the second Astronomer Royal, Edmond Halley, who is closely linked with its history.

In 1682 a bright comet appeared, and was observed by Halley. He worked out its orbit, and found that it moved strangely like other comets previously seen in 1607 and in 1531. Halley realized that the three bodies must in fact be different returns of the same comet, and he predicted that it would be seen again in 1758. Though he did not live to see the vindication of his prophesy, the comet was duly picked up on Christmas Night by a German amateur using a 6-inch telescope, and it actually passed perihelion on March 12, 1759, after which it vanished until the return of 1835. It was

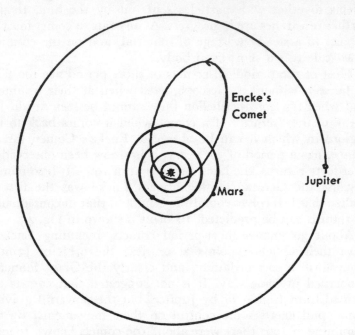

Fig. 45. Orbit of Encke's Comet.

seen once more in 1910, and is due back in 1986. It was Halley's Comet, too, that shone down on Saxon England in the early part of that most "memorable" year, 1066, and there are records of it that go back to before the time of Christ.

At present (1967), Halley's Comet lies very near the orbit of Neptune, and it is interesting to see what will happen to it in future years. The first thing shown from Fig. 46 is that the motion is retrograde, so that it is moving "the wrong way along a one-way street" in the same manner as Phœbe and the four outer moonlets of Jupiter. In 1948 it reached aphelion, and started to draw back slowly towards the Sun, crossing the mean path of Neptune in 1967. Another 10 years will bring it as close as Uranus, but after 1980 it will be moving so much more rapidly that it will cover the rest of the distance to perihelion in only another 6 years. By 1987 it will have receded

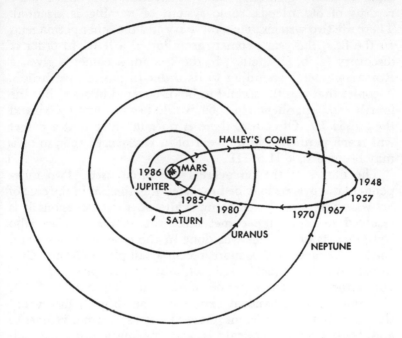

Fig. 46. Orbit of Halley's Comet. Aphelion—1948; perihelion—
1986.

once more beyond Jupiter, and unless we have by then completed more powerful telescopes we shall lose sight of it for another three-quarters of a century.

Unfortunately no other comet of reasonably short period can compare with Halley's, and most are faint telescopic objects, generally with tails that are either very faint or else absent altogether. One or two comets have peculiar orbits, a good example being known by the cumbersome name of Schwassmann-Wachmann I. Here the orbit lies entirely between those of Jupiter and Saturn, and is comparatively circular, so that the comet can be observed at any time when conditions are favourable. It is usually very faint, but sometimes brightens up considerably, so that it is a worth-while object for observers equipped with large telescopes.

Since four or five new comets are discovered every year, some of them genuinely new discoveries and others mere returns of old friends, some system of naming is essential. There are two systems, one temporary and the other permanent. In the first, the year's comets are allotted a letter in order of discovery (a, b, c, d, etc.); in the second, a comet is given a Roman numeral according to its order in passing perihelion. A comet that was the second to be discovered in 1962, but the fourth to pass perihelion in 1962, would become first 1962 b and then 1962 IV. Of course, there is no guarantee that a comet will reach perihelion in the year of its discovery; 1965 m or n may become 1966 II or III.

The names of the discoverers are often used. Two independent discoverers may be bracketed together, as in the case of Schwassmann and Wachmann, while on rarer occasions it is resolved to name the comet after the mathematician who computes its orbit. This was done in the case of Halley's and Encke's Comets, and a more recent example is Crommelin's Comet, which can just be seen without a telescope at a favourable return, and has a period of 28 years. The late A. C. D. Crommelin, a well-known expert on the subject, discovered that comets seen at different returns by Pons, Coggia, Winnecke and Forbes were identical. It was obviously just to attach Crommelin's name to it rather than to retain the names of all four discoverers.

Not all comets are periodic. Some have orbits which are almost parabolic (open curves) (Fig. 47), so that after having passed perihelion they retreat into space, never to return. There is little obvious difference between an open curve and a very long ellipse, and there are many comets which have periods of such length that they will not be seen again for generations. For instance, Quénisset's Comet of 1911 has an estimated period of over 9,000 years, and if this figure is accepted the modern return was the first since the end of the last Ice Age. Comet 1902 III seems to have a period of over a million years. It must be stressed, however, that periods of this kind are quite unreliable, and all we can say with

142

certainty is that the periods of such comets are extremely long.

It used to be thought that comets with open orbits came from space, visited the Sun once, and returned to interstellar space. This is not now believed to be the case. Comets are insubstantial bodies, at the mercy of the planets, so that their orbits may at any time be violently perturbed, and the results are sometimes remarkable. Jupiter, owing to its tremendous mass, has the greatest influence.

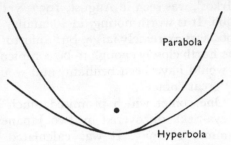

Fig. 47. Open curves; parabola and hyperbola.

Apart from Halley's, most great comets are either non-periodical or else have periods of hundreds of years. Several were seen during the nineteenth century, but since 1910 there has been a relative dearth of them, and by the law of averages one is very much overdue. There have of course been comets visible without telescopic aid, but faint objects at the limit of naked-eye visibility are very different from the spectacular Great Comets of the past. The comet of 1843 had a tail which stretched right across the sky, while that of 1858 (Donati's) was peculiarly beautiful in view of its triple tail, curved like a scimitar. Let us hope that the rest of our century will be less barren.

Much the most interesting comet of recent times was Arend-Roland, discovered in November 1956 by the two Belgian astronomers after whom it was named. Though hardly a "great comet" in the true sense of the word, it was a conspicuous naked-eye object for a short time in April 1957. On the 27th of that month, I had a particularly good view of it in the late evening; the nucleus lay close to the star Alpha Persei, and the long tail extended upwards from the horizon, so that it was a splendid sight in binoculars or a low-power telescope (to me, it seemed most impressive with binoculars). One of the interesting features of this comet was a curious "reverse tail". This "reverse tail" was still faintly visible on April 27, and is shown in Plate XI, while the nucleus of the comet then appeared almost stellar. Arend-Roland is a "non-periodical" comet; on April 20, 1957, it reached its minimum distance from the Earth, slightly over 50 million miles. Another naked-eye comet, Mrkos', was seen in August, 1957; yet another, Burnham's, in 1960. It is worth noting, incidentally, that Pereyra's Comet of 1963 was extremely large, but unfortunately never approached the Earth closely enough to be conspicuous. Had it come nearer it would have been brilliant, and would have been classed as a "great comet".

One comet which promised much but achieved little was Ikeya-Seki, discovered by two Japanese amateurs in the late summer of 1965. It was calculated that at perihelion, on October 21, the magnitude would be about −9; the comet was a "sun-grazer", similar to the great comet of 1882, and it was thought that there would be a long, glorious tail. On the morning of October 21 there was considerable interest everywhere. Personally, I went up in an aircraft over Ireland—by courtesy of the Royal Air Force, who provided a Shackleton aircraft for my benefit. We duly went up, taking off at 5.0 in the morning; with me was H. Grossie, of Armagh Observatory, and we carried photographic equipment, hoping to obtain some useful pictures above the cloud-level. Conditions were good, but the comet was conspicuous only by its absence, and we landed at 8.15, after sunrise, feeling rather disgruntled. We then learned that the Cambridge astronomer, D. W. Dewhirst, at 20,000 feet over East Anglia, had been similarly unsuccessful, and there were no reliable reports that the comet had been seen anywhere over Britain. (There was the usual crop of odd reports, and one daily paper published an impressive picture of what was said to be the comet, but was actually a cloud!)

What had happened, apparently, was that the comet had been reduced in size during its perihelion passage, when it had been within a few hundred thousand miles of the Sun, so that it emerged in a much fainter condition. Certainly it proved to be a great disappointment. Soon after perihelion it was seen from the Lowell Observatory in Arizona, from which the altitude above the horizon was much better than from Britain, and it did look imposing, even if it did not attain its expected grandeur. There is no telling how long we must now wait before seeing a new Great Comet comparable with those of the past.

When a periodical comet is due to return, its expected position at the time of anticipated recovery is given in a yearly publication such as the B.A.A. *Handbook*. The positions given are usually accurate enough for quick identification. Last time Encke's Comet came round, I picked it up without difficulty as soon as it came within range of my portable 3-inch refractor; and I am not, and never have been, a regular observer of comets.

The chief scope from the amateur's point of view is that many comets appear unexpectedly, and completely "out of the blue". There is always the chance of making a discovery, and some amateurs are adept at it, so that they have established international reputations.

Comet-sweeping is therefore a worth-while occupation, but the beginner must resign himself to many disappointments and many hours of fruitless searching. He may not discover a comet for years, or he may never discover one at all. There is however a great consolation, since even if he fails to find a comet he will be certain to come across many stellar objects of real interest.

Never use a high magnification. What is needed is a large field of view, and in any case a powerful eyepiece is of little use upon a badly-defined, fuzzy object such as the average comet. However, a small telescope will not collect enough light, and probably a 6-inch is the minimum aperture to hold out any real hope of success. The magnification should be in the region of 30 to 50.

Having selected the region to be swept, the telescope is moved slowly along in a horizontal direction (if on an altazimuth mount), with the observer keeping a careful watch all the time. Stars, clusters and other objects will creep through the field, and the slightest relaxation of attention may mean that a vital comet is missed. At the end of the sweep the telescope is raised or lowered very slightly, and an overlapping sweep taken in the opposite direction. After this process has been carried on until the whole area has been covered, it should be repeated several times until the watcher is satisfied that no dim, misty object can have escaped him.

Much patience is called for, and things are made more

difficult by the presence of star-clusters and nebulæ, which look very much like comets. The name "star-cluster" speaks for itself, while a "nebula" is rather similar in appearance, and is made up either of stars or of gas. If you are sweeping the heavens in search of a comet, and happen to find a misty object that is certainly not a star, it is unwise to jump to any conclusions. Reference to an atlas will probably show that the object is a cluster or a nebula that has been known for centuries.*

There is an interesting story about these clusters and nebulæ. Charles Messier, a famous comet-hunter of the eighteenth century, was persistently misled by uncharted stellar objects, and eventually he drew up a catalogue of "objects to avoid", rather as a navigator charts shoals in a strait. Nowadays Messier's comets are forgotten by all but a few enthusiasts, while his catalogue of clusters and nebulæ remains the standard. Messier himself would undoubtedly have seen the irony of the situation.

Many comets will lie somewhere near the Sun's line of sight when they are approaching perihelion, and nearly all remain undiscovered until they are well within the orbit of Mars, particularly as the average comet brightens up considerably as it draws near the Sun and the heat acts upon the particles in the nucleus. Consequently, the most promising areas of the sky for sweeping, for an observer in the northern hemisphere, are the west and north-west after sunset and the east and north-east before sunrise. It is also worth sweeping in the low north. It is no use beginning until the sky is really dark, since a faint comet will be drowned by any background light.

Though more new comets will be seen in these directions than in others, there is no hard and fast rule. A comet may appear at any moment from any direction; it may have an open or a closed orbit, it may be highly inclined, it may be moving in a retrograde or wrong-way direction. Comets have been called the stray members of the Solar System. Flimsy, harmless and of negligible mass, they can do harm to nobody. Moreover, they are short-lived upon the astronomical time-

* Odd things have happened now and again. Not long ago, one observer reported a "comet" that proved to be merely a reflection in his telescope, and there have been other similar cases. Even better was the "discovery" of a bright red star in the constellation of the Bull, not far from Aldebaran. It turned out to be Mars.

146

scale, and several short-period comets seen at several returns during the past have now vanished for good. Such are the comets of Biela and Brorsen.

Apart from comet sweeping, the amateur who has equipped himself with an equatorial mount, a measuring device and perhaps a camera, can do valuable work in checking the positions of comets from night to night. Mathematically-minded workers may prefer to make a hobby of computing orbits. This is not an easy process, and real skill is needed, but anyone who has the necessary ability and patience will soon find that his services are in great demand—more especially if he is the owner of a computing machine!

The link between cometary and meteoric astronomy is perhaps shown most clearly by the interesting case of Biela's Comet, whose peculiar career caused many astronomers many sleepless nights. The comet was discovered by Biela, an Austrian astronomer, in 1826, and found to be identical with comets previously observed in 1772 and in 1805. It was one of Jupiter's short-period group, and had a period of about 6¾ years. It returned in 1832 as predicted, was missed in 1839 owing to its unfavourable position in the sky, and returned once more in 1845.

Up to then, the comet had behaved in a perfectly normal manner, but during the return of 1845–46 it astonished observers by splitting into two pieces. Where there had been one comet, twins could be seen, sometimes with a kind of filmy bridge between them. Sometimes the two were nearly equal, sometimes the original comet was the brighter. Both faded gradually into the distance, and the return of 1852 was eagerly awaited. This time the two comets were farther apart, the second following the first rather like a child following its mother. At the 1859 return conditions were again hopelessly bad, but in 1856–66 the comet should have been an easy telescopic object. Yet it was searched for in vain. There was no trace of it; so far as could be made out, Biela's Comet had disappeared from the Solar System. Comets have been nick-named "ghosts of space", but no ghost could possibly have done a more successful vanishing act.

The next return should have taken place in 1872. Again the comet was absent, but in its place appeared a rich shower of meteors. Coincidence can be ruled out, and for years afterwards meteors were seen each year at the time when the Earth crossed the path of the dead comet. This shower is still active about November 28 annually, though it has now become very feeble.

It would be misleading to say simply that Biela's Comet "broke up" into meteors. There is more in it than this, and the position has been made clearer by the associations of other comets with other meteor showers; Halley's Comet, for instance, is linked with the meteor shower seen each year during the first week in May, and known as the Aquarids. Débris must be spread widely along the track of a comet, though once again it would be misleading to suppose that all meteors must be connected with comets.

Most people have wild ideas about the sizes of the particles which become incandescent and are rapidly burned up to become shooting-stars. Actually, the particles are very small. A body the size of a grape would produce a brilliant fireball, while the average bright meteor is due to a particle about a tenth of an inch in diameter. Like comets, meteors are less important than they seem.

A meteor travels round the Sun in an elliptical orbit, sometimes as a member of a shoal ("shower meteor") or as a lone wolf ("sporadic meteor"). If it comes close to the Earth, and is moving in a suitable fashion, it may enter the upper atmosphere at a relative speed of up to 45 miles per second. Below an altitude of 120 miles or so, there is enough air to cause appreciable resistance; heat and visual radiation are generated, and the hapless meteor is generally destroyed, ending its journey in the form of fine dust. Millions of shooting-stars enter the Earth's atmosphere every day. Most are smaller than grains of sand; the so-called micro-meteorites, which have been investigated recently by means of sending high-altitude rockets above the densest layers of the atmosphere, seem to have diameters of something like 5/1000 of an inch, and may be similar to the particles which cause the glow of the Zodiacal Light.

Sporadic meteors may appear from anywhere at any time, but shower meteors are more obliging. If the Earth passes through an area in space which is rich in meteors, the ordinary laws of perspective will cause the meteors to appear to radiate from one point. This is shown in Fig. 48, where all the meteors appear to converge towards a distant point P, which can be regarded as the apparent "radiant" of the meteors.

A meteor shower is named according to the constellation in which the radiant seems to lie. For instance, one major shower visible each November has its radiant in Leo, the Lion, and is thus called the Leonid Shower; of course, this does not mean that all the meteors appear near Leo, but merely that if the paths were plotted back, they would converge to a small area in Leo known as the radiant. Similar, the October Orionids radiate from Orion, and the August Perseids from Perseus.

Some of the annual showers are more important than others, and a list is given in Appendix XIV, but really spectacular displays are very rare. Such were the showers of 1833 and 1866, when the Leonids (associated with Tempel's periodical comet) were much more numerous than usual, and it was said that shooting-stars seemed to "rain down like snowflakes".

In fact, the Leonids had had a long and spectacular history, and had been consistent in providing major displays every 33 years. After 1866, the next was due in 1899—but by then, unfortunately, the meteor swarm had been affected by planetary perturbations, and the main cluster missed the Earth, so that the expected display did not materialize. The next return was due in 1933 (not 1932), but again there was nothing of note.

Conditions seemed more promising for 1966. The Leonid displays of 1963 and 1964 showed an encouraging increase, and this was also true of 1965, though for that year the observations were hampered by the inconvenient presence of the full moon. Much was hoped for 1966, and earlier in the month I put out a television appeal for what is known officially as "audience participation". With me was H. B. Ridley, the Director of the Meteor Section of the British Astronomical Association. We announced that charts and 'answer cards' would be distributed, and during the next few days the B.B.C. dispatched more than 10,000 of these charts and cards to people who wrote in for them.

The result was a sad anti-climax. In Ireland, where I was observing, the skies were reasonably clear, but at an early stage it became evident that the Leonids were going to fail us yet again. We saw some meteors, and plotted a radiant, but the display was so poor that nobody would have noticed it except by careful, systematic watching. Matters were very different elsewhere. As seen from parts of the United States (Arizona, for instance) the hourly Leonid rate reached 100,000—it was the greatest display of the century. Maximum occurred at about 12 hours G.M.T., while it was daylight in Europe; in fact, British observers missed the display by six hours. Yet the counts made by British amateurs were valuable scientifically.

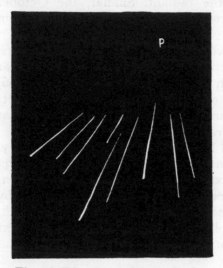

Fig. 48. Diagram to illustrate the principle of a meteor radiant. The meteors are assumed to be parallel, but to the observer the paths will seem to converge to the point P.

Another good display of the present century occurred on October 9, 1946. The shower was connected with the periodical comet Giacobini-Zinner, and for some hours meteors appeared at the rate of one per second. The main spectacle took place during daylight over Europe, so that it was seen only by observers in the New World.

To find out the speed, height and orbit of a meteor, three data must be provided: the point of appearance of the meteor, the point of disappearance, and the duration. Clearly it is necessary for the same meteor to be observed by two workers placed at least twenty miles apart (more if possible). A single observer cannot do much if he has to depend only upon his own labours.

No instruments are needed for meteor recording, but the observer has to have a really good knowledge of the constellations, as otherwise he will be unable to plot the track. The

150

track must be plotted on a star map, but it is unwise to look down as soon as the meteor has vanished and try to record where it went, since errors are certain to creep in. The solution is to check the path by holding up a rod or stick along the track where the meteor passed, which will give you the chance to take stock of the background and ensure that no mistake has been made. When you are satisfied, either draw the path on your chart or note the exact positions of the beginning and end of the track, and then write down: time of start, duration, duration of luminous trail, brightness (compared with that of a known star or stars), colour (if any), and any special features.

Meteor watching is a lengthy and often a cold business. Standing out for hours during a January or February night is enough to chill the enthusiasm of the hardiest observer. Nevertheless, until recently all researches were based upon the patient work of amateurs, among whom the name of W. F. Denning will always be remembered. It is fair to say "until recently", because in 1946 an entirely new method of recording was brought into operation, that of radar.

The passage of a meteor through the atmosphere has a pronounced effect upon the air-particles, and these effects can be detected by radar. Reduced to its barest terms, radar involves sending out an energy wave, and recording the echo as the wave is bounced back after hitting a solid object. A meteor trail is not of course a hard body, but it acts just as violently, and radar detection of shooting stars has now been in progress for some time. The method is unhampered by clouds or daylight, and it would be idle to pretend that it has not affected the value of amateur visual work, though the naked-eye watcher can still make himself useful.

Casual meteors are fairly frequent, and a watchful observer will seldom fail to record fewer than five or six per hour, but it is of course far more entertaining (though not necessarily more useful) to concentrate upon some definite shower. Occasionally there will be so many meteors in quick succession that the watcher will be hard pressed to record them all, but this will not happen often, and there must be long periods of patient waiting.

It is interesting to note that visual meteors are twice as abundant in the period from midnight to 6 a.m. as during the

151

period from 6 p.m. to midnight. In the evening, we are on the "rear" of the Earth as it moves in its orbit, so that visible meteors have to catch us up; in the morning hours we are in the "front" position, so that meteors meet us coming. More meteors are to be expected after midnight than before it, and obviously the morning meteors will have greater relative speed, just as a car moving at 30 m.p.h. and meeting a second car moving at 35 m.p.h. will be badly damaged if the collision is head-on, but only bumped if rammed from behind. It is the relative speed of a meteor which is the main factor in its brightness, so that the morning meteors will be more brilliant and hence easier to record.

Though meteors and comets are so unpredictable, at least when compared with the planets, studies of them are full of interest. Moreover, there is always the chance of making a spectacular discovery or discoveries—as happened to a British amateur, G. E. D. Alcock, in 1959, when he found two new comets in quick succession. On the other hand, the amateur who wishes to make a serious, useful study of meteors is more likely to concentrate upon photographic work; there is a great deal to be done, for instance, in photographic recording of meteor spectra. Many hours of exposure are needed to capture even one meteor spectrum, but a successful photograph is of great value. One of the pioneers in this field is H. B. Ridley, who, like Alcock, is an amateur.

Larger bodies, more nearly related to the asteroids than to ordinary meteors, survive the complete drop to the ground, and are known as meteorites. Most museums have collections of them, and an expert can soon tell what is meteoritic material and what is not, though the layman is easily misled. In general, meteorites are divided into two classes, stones (aerolites) and irons (siderites).

Large meteorites are rare. The biggest of modern times fell on June 30, 1908, and landed in Siberia, blowing pine-trees flat for 50 miles all round the impact point; there must have been many earlier falls—witness, for instance, the large crater in the Arizona Desert, which is undoubtedly due to a prehistoric meteorite impact. Luckily, the dangers to human life are so slight as to be negligible.

The last really interesting fall occurred in England on Christmas Eve, 1965. A meteorite flashed across the Midlands, attracting considerable attention, and broke up; fragments of it came down at Barwell, in Leicestershire. (I even found one myself when I visited the site some time later.) The original weight of the meteorite must have been about 200 pounds, which is a British record. One fragment went through the window of a house in Barwell, and was found later nestling comfortably in a vase of artificial flowers.

It is fitting to end this brief survey of the Solar System with the meteors and meteorites, its most insignificant members. We have described the Sun, the Moon, the planets and their satellites, the vivid glow of the aurora and the pale radiance of the Zodiacal Light, and the flimsy and unpredictable comets, so that there is variety in plenty; but even the Sun itself is a very junior member of the Galaxy, and we must keep our sense of proportion.

Though the amateur's greatest scope is with the bodies of the Solar System, and many stellar problems cannot be tackled without using complex equipment, it would be a mistake to confine ourselves only to the Sun's family. Greater problems remain to be studied, and in any case a knowledge of the stellar universe can give one endless enjoyment. We must remember Carlyle's lament: "Why did not somebody teach me the constellations, and make me at home in the starry heavens?"

Chapter Twelve

THE STELLAR HEAVENS

WHEN MEN OF ancient times looked up into a starlit sky, they could see many hundreds of tiny, twinkling points that seemed to be arranged in definite patterns. It was natural, then, for the stars to be grouped into definite "constellations", each named after a deity, a demigod or else some common object. Orion the Hunter, Hercules of legendary strength, and Perseus with the Gorgon's Head mingle with the Dragon, the Fishes and the Cup. Forty-eight separate constellations are listed in the great catalogue contained in Ptolemy's Almagest, and may therefore be said to date from the dawn of astronomy.

The names are generally used in their Latin forms, so that the Dragon is "Draco" and the Fishes "Pisces". Any amateur who means to do serious work in the field of stellar research should become accustomed to the Latin names, which are in any case easy to remember. A full list, with the English equivalents, is given in Appendix XV.

Ptolemy's 48 constellations are still used to-day, but others have been added since. Some of these new groups lie near the south celestial pole, so that they never rise in the latitude of Alexandria, and Ptolemy naturally knew nothing about them; others have been formed by taking pieces away from the original 48. Further proposed additions with barbarous names such as Sceptrum Brandenburgicum, Officina Typographica and Lochium Funis have been mercifully forgotten, though one of the rejected groups, Quadrans Muralis (the Mural Quadrant) has left a legacy in the form of the name of the annual Quadrantid meteor shower seen each year from January 3 to 5.

Probably the most famous of the constellations are Ursa Major (the Great Bear), Orion, and Crux Australis (the Southern Cross). Of these, Ursa Major lies in the far north of the sky, so that in England it never sets, while Orion is crossed by the celestial equator and Crux is so far south that it never rises in our latitudes. Stars which never set are termed "circumpolar", so that Ursa Major is circumpolar in England. To give a full explanation of the apparent movement of the star-sphere would be rather beyond our present scope, but something must be said about the essential terms Right Ascension and Declination. Broadly speaking, these are the celestial equivalents of longitude and latitude on the Earth's surface, though there are certain important differences in detail.

Declination is reckoned in degrees north or south of the celestial equator, while the equator itself is merely the projection of the Earth's equator in the sky. Clearly the north celestial pole will have declination 90 degrees north (+90°), and Polaris, the Pole Star, with its declination of greater than +89°, is so close to the polar point that it always indicates the approximate north pole. Observers in the southern hemisphere are not so lucky, since there is no bright star placed conveniently at the south polar point.

To anyone observing from the north pole of the Earth's surface, Polaris would appear to remain virtually overhead; its altitude above the horizon would be greater than 89°. At Greenwich (latitude N. 51½°) the altitude of Polaris is 51½°; on the equator (latitude 0°) Polaris has of course no altitude at all—in other words, it lies right on the horizon. South of the terrestrial equator, Polaris never rises, so that it will never be seen.

The point at which the Sun crosses the celestial equator in its springtime journey from south to north is known as the Vernal Equinox, or First Point of Aries. The Sun reaches this point about March 21 each year, and crosses the equator once more, this time from north to south, six months later at the Autumnal Equinox, or First Point of Libra. The vernal equinox is to the sky what the Prime Meridian is to the Earth, since all positions are reckoned from it; but we must remember that it is a point of definite significance, whereas Greenwich was chosen as a standard for longitude merely because the famous Observatory happened to have been built there.

The angular distance of a star eastwards of the Vernal Equinox is known as the star's right ascension. It can be given in degrees, but is more usually measured in hours, minutes and seconds, because such a method is more convenient.

To explain this, we must refer to the "meridian" of any observing point on the Earth, which is the great circle on the star-sphere passing through both celestial poles and also through the overhead point (the "zenith") of the place of observation. Clearly, a star on the "meridian" will be at its maximum height above the horizon. The First Point of Aries must pass across the meridian at any place once every 24 hours (sidereal time), and the difference between this time and the time of the star's meridian passage will give us the right ascension of the star. For instance, Sirius reaches the meridian 6 hours 43 minutes after the First Point of Aries; therefore, the right ascension of Sirius in the sky is 6 hours 43 minutes.

The slight shift of the celestial pole, described in Chapter II, means that a star's right ascension and declination alter very gradually over the years. In this book (and in most modern star atlases) the positions are given for the year 1950, since it will be a long time before the error becomes great enough to be at all worrying. As a matter of interest, the First Point of Aries has shifted so much since olden times that it has moved out of Aries altogether, and now lies in the neighbouring constellation of Pisces, the Fishes.

A telescope equipped with setting circles and clock drive can be swung to any desired right ascension and declination, so that as soon as the position of a body is known the telescope can be directed straight towards it, without bothering about searching. Since the planets can be found in the same way, this is much the easiest method for picking up Mercury or Venus in broad daylight. It is clear, of course, that while the right ascension and declination of a star will remain virtually constant, those of the Sun, Moon and planets will alter appreciably over a very short period.

Dividing the stars into constellations, and naming the brightest objects, is enough for a rough classification. Most of the leading stars have proper names, such as Sirius, Canopus, Rigel, Vega and Capella. On the other hand it would be a hopeless task to allot special names to each star, and we have recourse to letters or numbers.

A method used by Bayer, who drew up a famous star catalogue in 1603, has stood the test of time so well that it will certainly never be altered. On this system, each of the leading stars of a constellation is allotted a Greek letter, beginning with Alpha for the brightest object and ending with Omega for the faintest. In Aries the Ram, for instance, the brightest star is Alpha Arietis (Alpha of the Ram), the second brightest Beta Arietis, and the third brightest Gamma Arietis. Unfortunately the strict order is often not followed, so that the system has become rather chaotic. In Orion, Beta is the brightest star, followed by Alpha, Gamma, Epsilon, Zeta, and then Kappa, with Delta an "also ran". A list of the Greek letters, with their English names, is given in Appendix XIX.

This is all very well, but it can deal only with the 24 principal stars in each constellation, which in some cases (such as Orion) is not nearly enough. Flamsteed, the first Astronomer Royal, preferred to give the stars numbers, beginning in each constellation with the star of least right ascension. Still fainter stars, not listed by Flamsteed, have been allotted numbers according to later catalogues, and the result is that each bright star has several designations; Rigel in Orion is known also as Beta Orionis and as 19 Orionis. As time goes on, the proper names of the stars are becoming less and less used, with the Greek letters and the numbers taking their places.

It is also necessary to have some scale of reckoning apparent brilliancy. This is done by classification into "magnitudes" but the scale sometimes causes confusion, since the lowest values indicate the most brilliant objects. Bright stars are of magnitude 1, and the faintest visible to normal eyes without a telescope are of magnitude 6, while with powerful telescopes stars down to the 21st magnitude can be detected. Modern instruments known as "photometers" can measure the brightness of a star very exactly, and in catalogues the value is given to 1/100 of a magnitude. Polaris, for instance, is of magnitude 1·99, so that it may be regarded as a standard star of the second magnitude.*

A few stars are actually brighter than magnitude 1·0, so that they have values of less than unity; examples are Rigel (0·15) and Altair (0·80), Rigel being appreciably the brighter of the two. Four stars—Sirius, Canopus, Alpha Centauri and Arcturus—have minus magnitudes. On the stellar

* The magnitude of Polaris is very slightly variable.

scale, Venus at its brightest is of magnitude $-4\frac{1}{2}$, while the Sun is about -27. The magnitude scale is based upon a definite mathematical ratio, but this need not concern us at the moment.

The stars are of different luminosities, and are at different distances from us, so that our constellation groups are due to mere line of sight effects. In Ursa Major, for instance, one of the seven bright stars (Alkaid) is much more remote than the other six, while Polaris, in Ursa Minor or the Little Bear, is twice as distant as Alkaid. Merely because two stars are in the same constellation, we need not suppose that they have any

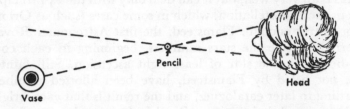

Fig. 49. Diagram to illustrate the principle of parallax.

connection with each other. There is an easy way of showing this. If you look at a gatepost as seen against the background of a clump of trees, you do not suppose that the gatepost has any real connection with the trees.

As described by our scale model on page 30, the stars are so remote that their distances are not easy to measure. The first reliable results were obtained by using the method of "parallax", which is interesting enough to be explained more fully, even though it is useless for any but the very nearest stars.

The best way to demonstrate parallax is to make a practical experiment with a pencil, holding it up in front of your face and looking at it with alternate eyes. First align the pencil with some object, such as a vase on the mantelpiece, using your right eye only. Now shut your right eye and open your left, keeping the pencil still. The pencil will no longer seem to be in line with the vase; it will seem to have shifted. If you know the distance between your eyes, and the angular amount by which the pencil appears to have shifted, you can work out the actual distance of the pencil by using fairly simple mathematics (Fig. 49). This apparent shift in position is a measure of the pencil's parallax.

Much the same principle can be used to measure the distance of a relatively near star seen against a background of more distant stars, but the base-line used has to be enormously long. Fortunately Nature gives us such a base-line; the Earth swings from one side of the Sun to the other in a period of six months, shifting 186 million miles in position (Fig. 50). If S is our "near" star, it will appear to be at position S1 in January, but at S2 in July, so that if we measure the angular shift we can find the distance. (The diagram given here is hopelessly over-simplified and out of scale.) The actual amount of the shift is so minute that it is hard to measure, while there are numerous corrections to be made. However, there is nothing complex in the basic principle of the method, and it was in this way that Bessel managed to measure the distance of the fifth-magnitude star 61 Cygni, in 1838.

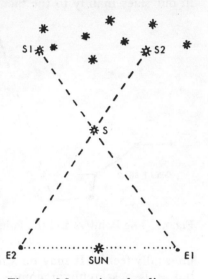

Fig. 50. Measuring the distance of a star by parallax. E1, E2—Earth; S—star; S1, S2—apparent positions of S with respect to more distant stars. The diagram is of course grossly out of scale.

The parallax method breaks down altogether for all but the closer stars, because the shifts become too small to be properly measured. At 160 light-years the method has become untrustworthy, and at 600 light-years it is quite useless. Indirect methods have had to be developed, and most of these involve finding out the actual luminosity of a star as compared with the Sun, since as soon as we know the real brilliance and the apparent brilliance we can find the distance—much as we can judge the distance of a lighthouse if we know the power of the lamp and can measure how bright it appears to us.

Even the nearest of all the stars, Proxima Centauri in the southern sky, is immensely remote, so that in comparison even

Pluto is very close at hand. The distance in miles is about 25 million million, or 4⅛ light-years.

Sirius, which appears the brightest star in the sky, is 26 times as luminous as the Sun, but it owes its supreme position in our skies mainly to the fact that it is relatively close to us, since it lies at a distance of only 8½ light-years. Canopus, in the southern constellation of Argo Navis (the Ship Argo), looks only a little less bright than Sirius, but is at least 650 light-years away instead of 8, so that it is clearly much more luminous. It would in fact take 80,000 Suns to equal Canopus.

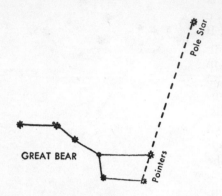

Fig. 51. The Pointers and the Pole Star.

However, we must not imagine that our Sun is unusually feeble. It may be a firefly compared with Canopus, but it is a searchlight compared with some of the dimmest members of the stellar system. The faint red star known as Wolf 359 has a luminosity of only 1/50,000 of that of the Sun, so that we need not be too humble. If anything, the Sun is rather above the average in brilliancy, though there is really no such thing as an "average star".

Just as the stars are of different distances and luminosities, so they are of different sizes, temperatures and colours. A glance at Orion will show that of the two apparently brightest stars, one (Betelgeux*) is orange-red, while the other (Rigel) is white or slightly bluish. Betelgeux is the larger of the two, but it is much the less luminous, and its surface is cooler than that of Rigel. In fact, the stars present an almost infinite variety, so that it is rare indeed to find two which seem to be exactly alike.

To the ordinary observer, the stars appear to remain in fixed positions. Two of the stars in the Great Bear, Dubhe and Merak, always point to the Pole Star (Fig. 51); they have done

* This name may be spelled in various ways, such as Betelgeuse and Betelgeuze. In using a final x, I have followed the advice of Arabic scholars.

(*Left*) Henry Brinton's 12-in. reflector at Selsey, Sussex. This has a skeleton tube, with rotatable head, and is equatorially mounted, with an electric clock-drive. The mounting is of the so-called German type, with the weight of the telescope tube balanced by a counterweight—or, in this instance, several counterweights.

(*Right*) Patrick Moore's $12\frac{1}{2}$-in. reflector at East Grinstead. Here the tube is solid, and the mounting is altazimuth, with manual slow motions. The telescope is of short focal length (72 in.) and is very easy to operate, but the need for constant adjustment is of course a serious handicap.

(*Left*) The rotatable head of Henry Brinton's reflector. The entire head can be moved round, so that the eyepiece can always be brought into a position convenient for viewing. This is a great advantage in view of the type of mount of the telescope; if the head did not rotate, things could become very difficult.

II. *Observatories*

(*Above*) J. Hedley Robinson's observatories at Teignmouth. The dome houses a 10-in. reflector; the entire building revolves. The run-off shed houses a 3·75-in. equatorial refractor.

(*Right*) Patrick Moore's observatory at East Grinstead. The upper part of the dome revolves. The observatory houses an 8½-in. equatorial, clock-driven reflector. It has now been transferred to Armagh.

One of F. W. Hyde's radio telescopes at Clacton, used for studying the Sun.

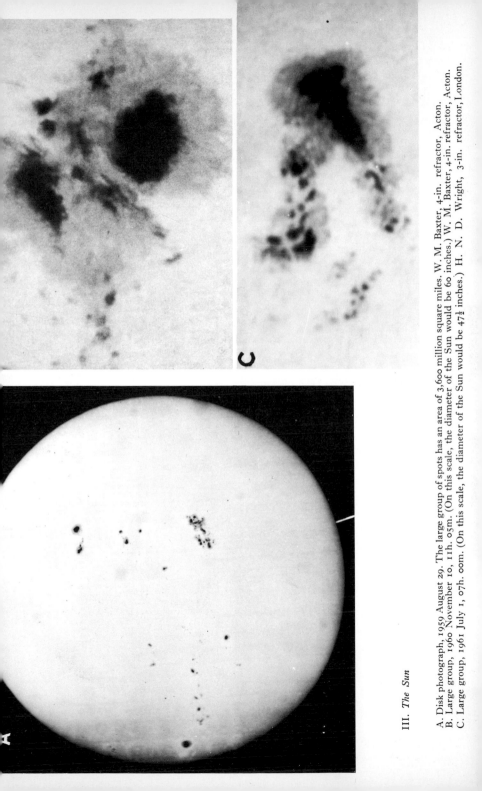

III. *The Sun*

A. Disk photograph, 1959 August 29. The large group of spots has an area of 3,600 million square miles. W. M. Baxter, 4-in. refractor, Acton.
B. Large group, 1960 November 10, 11h. 05m. (On this scale, the diameter of the Sun would be 60 inches.) W. M. Baxter, 4-in. refractor, Acton.
C. Large group, 1961 July 1, 07h. 00m. (On this scale, the diameter of the Sun would be 47½ inches.) H. N. D. Wright, 3-in. refractor, London.

A. *Mare Imbrium.* F. L. Jackson, 11·75-in. reflector, 1958 December 20, 20h. 45m. This is a picture showing a wide area; Plato is the dark-floored crater in the lower part of the Moon. In this and all other photographs, south is at the top and west to the left.

B. *Ptolemæus to Walter.* G. A. Hole, 24-in. reflector. This shows two great chains, Ptolemæus-Alphonsus-Arzachel and Walter-Regiomontanus-Purbach. The Straight Wall is seen as a bright line to the east (right), rather above the centre of the photograph.

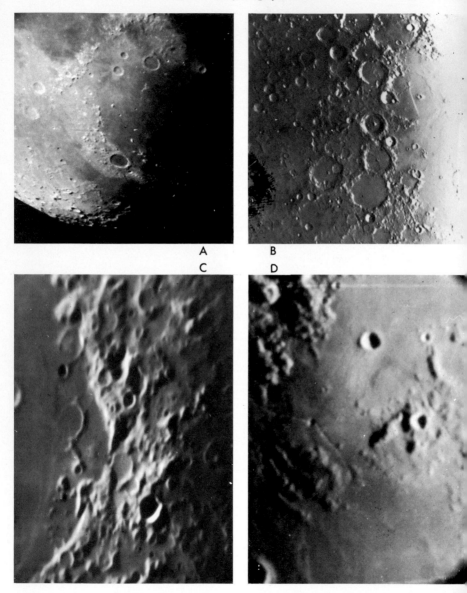

A B

C D

C. *East Part of Mare Crisium.* H. E. Dall, 15¼-in. reflector, 1961 February 19, 16h. This large-scale photograph shows fine details on the boundary of the Mare. Picard, on the Mare, is partly shown to the west (left).

D. *Hyginus Cleft Area.* G. A. Hole, 24-in. reflector. The Cleft is well shown, as is the surrounding area. The large scale of the photograph shows that the Cleft contains many crater-like enlargements along its length. The prominent crater to the south is Triesnecker.

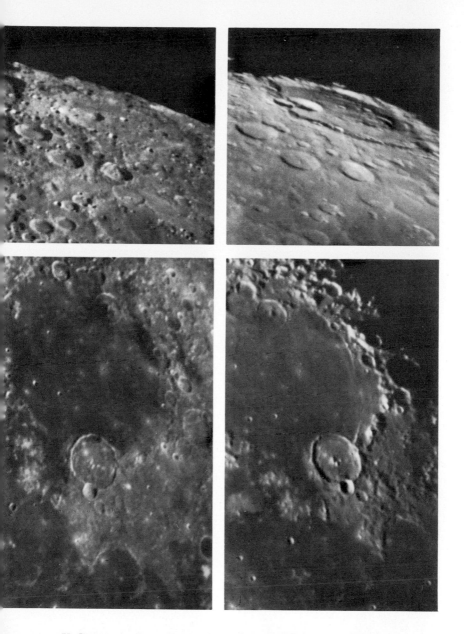

V. *Comparative Lunar Photographs by H. R. Hatfield, ($12\frac{1}{2}$ in. reflector)*

Upper: Bailly. (*Left*) 1966 October 6, 05.16. Bailly is well inside the visible disk. (*Right*) 1966 March 5, 22.59. Here, Bailly is on the terminator, and appears very prominently.
Lower: Mare Humorum. (*Left*) 1966 August 9, 03.17. The area is under fairly high light, with Gassendi well shown. (*Right*) 1966 March 3, 19.58. Here the Mare is close to the terminator, and some shadow can be seen inside Gassendi; the mountainous border of the eastern Mare Humorum is seen to advantage.

Photographs by T. W. Rackham, 6-in. reflector.

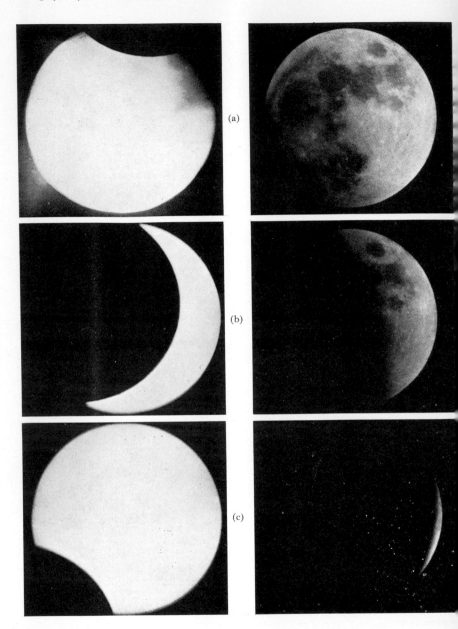

(a)

(b)

(c)

(*left*) Eclipse of the Sun, 1954 June 30, partial at Cambridge: (a) 11h. 20m. (b) 12h. 36m. (c) 13h. 37m. Some clouds can be seen in the first view.

(*right*) Eclipse of the Moon, January 19, 1954, total at Cambridge: (a) 0h. 50m. (b) 1h. 32m. (c) 1h. 55m.

VII. *Drawings of the Planets*

A. *Mars*, 1965 March 9, 01.25. 12½ in. reflector × 460. Patrick Moore. The Syrtis Major is shown to the upper left.

B. *Jupiter*, 1963 December 7, 16.53. 8½ in. reflector × 300. Patrick Moore. The Red Spot is shown, but the Hollow was not visible. Note that the two Equatorial Belts have merged into a continuous dark strip.

C. *Jupiter*, 1964 August 23, 03.10. 8½ in. reflector × 274. Paul Doherty. The Equatorial Zone is still dark; the Red Spot is shown, with a white spot south and preceding it.

D. *Jupiter*, 1966 October 11, 04.14. 10 in. refractor × 350. T. J. C. A. Moseley. The Equatorial Belts are now separate, and the Red Spot is somewhat inclined in its Hollow.

E. *Saturn*, 1963 August 7. 00.45. 8½ in. reflector × 300. Paul Doherty. The rings are well shown, with the Cassini Division.

F. *Saturn*, 1966 August 26, 23.30. 8½ in. reflector × 274. Paul Doherty. The rings are almost closed, but the shadow on the disk is prominent. Titan is seen some way from Saturn; the black spot on the planet's disk is the shadow of Titan.

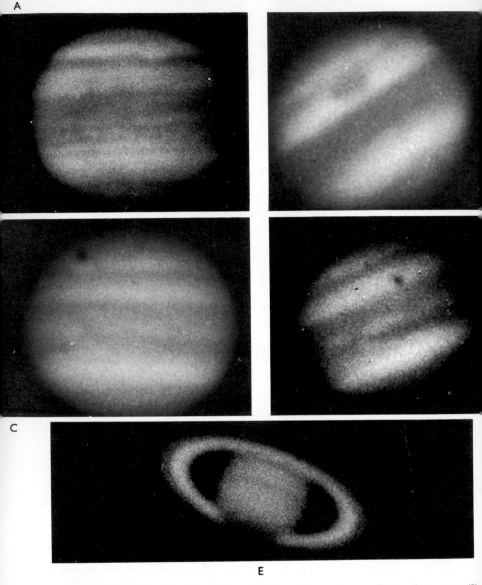

A. *Jupiter*. H. E. Dall, 15½-in. Cassegrain. 1963 November 2, 21h. 29m. Long. of c.m.: 340 (I) 297 (II).

B. *Jupiter*. W. Rippengale, 1962 July 30; the best of a series taken from 01h. to 03h. The Red Spot is beautifully shown.

C. *Jupiter, showing Transit of Ganymede*. W. Rippengale, 1963 September 28, 23h. 10m.

D. *Jupiter, showing Shadow Transit of Io*. W. Rippengale, 1963 October 27, 21h. 50m. Conditions were misty; the image was steady, but a long exposure needed (1½ to 2 sec.).

E. *Saturn in 1957*: H. E. Dall, 15½-in. Cassegrain. The rings were widely displayed; the main belt on the disk is shown, as well as the Cassini Division in the rings.

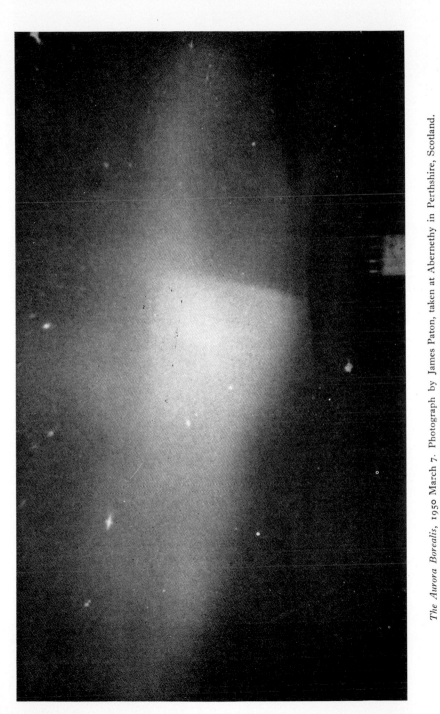

The Aurora Borealis, 1950 March 7. Photograph by James Paton, taken at Abernethy in Perthshire, Scotland.

X. *Comet Arend-Roland* photographed in 1957 at Peterborough by W. A. Granger (assisted by Mrs. Granger).

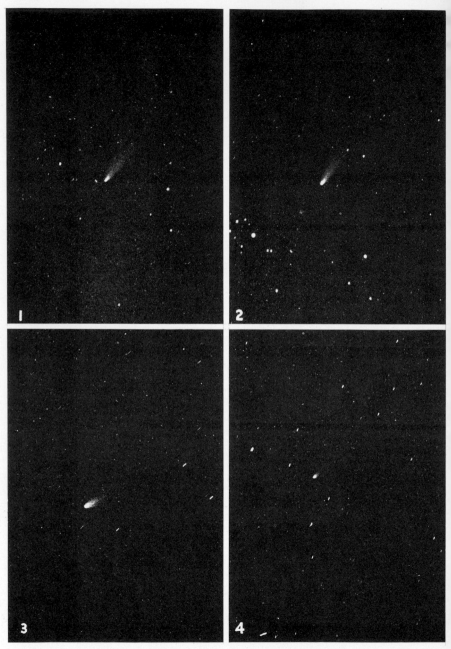

(1) April 27, 20.57-21.27; exposure 30 min. Position of comet: R.A. 3h. 10m., decl. +57°0′. 7-in. Aero Ektar, HP3.

(2) April 28, 21.04-21.40; exposure 36 min. R.A. 3h. 25m., decl. +52° 55′. 7-in. Aero Ektar, Zenith Astron. Plate.

(3) April 28, 21.04-21.40; exposure 36 min. 20-in. Aero lens, HP3.

(4) May 5, 21.44-22.24; exposure 40 min. R.A. 4h. 55m., decl. +61° 20′. 20-in. Aero Camera, HP3.

XI. *Comet Arend-Roland*, 1957 April 27. Photograph by Frank J. Acfield, Forest Hall. The unusual "spike" is well shown.

XII. *The Pleiades*. Photograph by R. E. Roberts, 9-in. reflector, exposure 30 min.

XIII. *Venus near the Pleiades*.   F. J. Acfield,
Forest Hall, 1956 April 2. Exposure 10 minutes.
f/5.8 camera.

This photograph by Howard Miles shows the track of 1960 Iota 1 (Echo 1 Satellite) as it passed through the constellation of Orion on March 1, 1964 at 21.55 hrs. The fading on the lower left side of the photograph is due to the satellite entering the Earth's shadow. The two brightest star-trails are those of Betelgeux and Bellatrix.

XV. *The Moon: Western Half.* The various formations may be identified on the key map p. 224.

XVI. *The Moon: Eastern Half.* The various formations may be identified on the key map p. 227.

so for generations, and will continue to do so for generations more. Of course, the old term "fixed stars" is misleading. The stars are moving about at high speeds, but they are so remote that it takes centuries for bright naked-eye stars to show obvious shifts in position, while the tiny annual shifts due to parallax can be detected only with the most refined instruments. Over the ages, however, the shifts will mount up, and eventually the two Pointers will no longer seem to line up with Polaris.

The slow movement of a star across the background is known as the star's Proper Motion, and must not be confused with the minute movement due to parallax. There is also a motion in the line of sight, termed Radial Motion (Fig. 52). If a star is coming straight towards us or away from us, it will have no proper motion at all, and will appear to remain still even over the lapse of centuries, but its radial

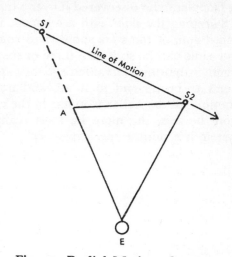

Fig. 52. Radial Motion of a star. S_1S_2 = actual motion, S_1A = radial motion. AS_2 = proper motion.

motion will be detectable by means of the spectroscope.

Since the Sun is an ordinary star, the other stars show spectra of much the same kind. Temperature differences and other factors will cause complications, but usually there will be the continuous rainbow crossed by dark absorption lines due to gases in the star's reversing layer (page 64). If the star is approaching us, the dark lines will be shifted slightly towards the violet or short-wave end of the spectrum, while if the star is receding the shift will be towards the red. By measuring the amount of the shift, we can work out the radial velocity of the star.

There is an everyday analogy to this. When a train whistles, the whistle is high-pitched so long as the train is coming towards us, because more sound-waves are entering our ears, per second, than would be the case were the train standing still. After the train has passed by, and begins to draw away, fewer sound-waves will reach us per second, so that the pitch of the whistle drops. Light can be regarded as a wave-motion, and when the source of light is moving away the "pitch" is shifted towards the long-wave or red end of the spectrum. This is known as the Doppler Effect, in honour of the Austrian physicist Doppler, who discovered it over a hundred years ago.

Sweeping the skies with a telescope is a fascinating occupation. Some of the stars show vivid colours; some are double, and some can be split into three or more components, so close together that to the unaided eye they appear as one star. There seems to be no end to it all, and no observer can hope to examine all the stellar wonders in the course of a lifetime. The more he sees, the more he must realize that our own Solar System is a minute speck in space.

Chapter Thirteen

THE NATURE OF A STAR

NEARLY EVERYONE who uses a telescope for the first time expects to see a bright star, such as Sirius or Rigel, enlarged to a massive globe filling half the field of view. Disappointingly, however, nothing of the kind is visible. If the disk of the star is of appreciable size, there is something wrong with the telescope —since not even the Palomar 200-inch reflector can show a truly measurable disk to any star.

This is not because the stars are small. Some of them are in fact big enough to hold the whole orbit of the Earth. The small apparent size is due to the fact that the stars are inconceivably remote. No amount of magnification upon our modern telescopes can improve matters, and if we want to study the stars themselves we must resort to indirect methods.

At first sight, therefore, it would seem as though we could never gain much information. But though the telescope is not by itself particularly helpful, it can be combined with the spectroscope to make a powerful weapon which can be used not only to analyse the materials which make up the stars, but also to investigate the inner regions, the "power-houses" where stellar energy is generated.

Most stars show spectra which are basically similar to that of the Sun (see page 64), but there are wide variations in detail. Over 90 years ago Father Secchi, one of the great pioneers in this field, found that there were well-marked "spectral types"; for instance, stars like Sirius showed prominent dark absorption lines due to hydrogen gas, while in the case of Rigel lines due to helium were dominant. Secchi divided the stars into four definite groups. A more comprehensive system, originated by E. C. Pickering (brother of W. H. Pickering, the lunar and planetary observer) increased the number to eleven, merging gradually into each other.

On this latter system, each type is denoted by a letter of the alphabet. It was originally intended to take the usual sequence of letters, but some of the early classes were found to be unnecessary—there is now no recognized Type C, for instance—and the series became muddled, until to-day it reads: W, O, B, A, F, G, K, M, R, N, S. Some thoughtful astronomer has invented the mnemonic "Wow! Oh, Be A Fine Girl, Kiss Me Right Now Sweetie", which is at least a good way to remember the correct series.

To describe the features of each type would require many pages, but it will be of interest to give a brief outline. The series given above denotes an order of decreasing surface temperature, W and O stars being the hottest and R, N and S the coolest; the Sun, as befits its undistinguished character, comes in Type G, somewhere near the middle of the list. A refinement is to divide each type into sub-grades, from nought to nine, so that A5 is midway between Ao and Fo.

Some W and O stars, known as Wolf-Rayet stars in honour of the two astronomers who first described them in detail, have surface temperatures of over 35,000 degrees Centigrade, so that they are the hottest of the normal stars. Their spectra are peculiar, having in some cases a large proportion of bright lines instead of the usual dark ones, and they have set astronomers many problems, some of which remain to be solved. Most Wolf-Rayet stars are very remote, so that they appear faint in spite of their great luminosity, though two of them (Zeta and Gamma Argûs) are of the second magnitude; Gamma is too far south to rise in England.

Rigel in Orion has a B-class spectrum, and in fact all the leading stars in Orion are of this type, with the obvious exception of Betelgeux. The surface temperatures are in the region of 25,000 degrees Centigrade, so that B-stars are highly luminous. Somewhat less hot are the A-stars such as Sirius, with temperatures of about 11,000 degrees Centigrade; stars of type F, such as Canopus, are cooler still, so that they appear yellowish. Hydrogen and helium lines are less conspicuous, but calcium vapour is much in evidence.

The Sun is a typical G-type star, with a surface temperature of 6,000 degrees Centigrade. Here, of course, our investigations are helped by the fact that the solar spectrum can be studied in great detail. Another good example of a G-star is Capella, which appears as one of the most conspicuous stars in our skies.

The remaining types are orange (K) or orange-red (M, R, N and S), with temperatures ranging from 4,200 degrees down to only 2,000 degrees. Types N, R and S are comparatively rare, and most of them are variable in brightness, while their spectra are complex and not at all easy to interpret. Arcturus in Boötes is of Type K, while Betelgeux, Mira in Cetus, and Antares in Scorpio belong to Type M.

It may be convenient to group the stars in this way, but we have only touched the fringe of the problem. Consider, for instance, two M-type stars, the brilliant Betelgeux and the dim Wolf 359. Betelgeux shines as brightly as 1,200 Suns, while Wolf 359 is a feeble body with only 1/50,000 of the Sun's candle-power, so have we any reason to class them together in the spectrum sequence? To say the least of it, they are ill-assorted companions.

One of the great discoveries of the early twentieth century was that apart from types W, O, B and A, the spectral classes tend to be separated into "giants" and "dwarfs". We can find many M-giants like Betelgeux, and many M-dwarfs like Wolf 359, but M-stars of intermediate luminosity are virtually absent. When it became possible to estimate the diameters of the stars, the distinction between giants and dwarfs became even more evident. Betelgeux is a vast globe about 200 million miles across, whereas Wolf 359 has a diameter of less than a million miles. If we picture a scale model and make Betelgeux a globe with a diameter equal to that of a cricket pitch, Wolf 359 will be represented by a croquet ball.

The discovery of the giant and dwarf divisions was followed by a very simple, straightforward theory about the life-history of a star. It was assumed that in its early life, soon after it condensed out of the interstellar dust and gas, a star was hardly hot enough to emit visible light. Naturally, it would tend to shrink, because the force of gravity would tend to pull all its matter together; this would cause heat, so that the star would become a large Red Giant like Betelgeux. As the shrinking went on, the star would become an Orange Giant (type K) and then a Yellow Giant (type F), before turning into a smaller but very hot Wolf-Rayet or B star. As would be expected on this theory, the most luminous white types are not divided into giants and dwarfs.

This would be the peak of a star's career. It would go on shrinking, but it would also become cooler, since its main energy would have been spent. It would pass down the dwarf series or "Main Sequence", becoming first an F-dwarf, then a G-dwarf like the Sun, and then a small, red star of one of the later types, finally losing all its heat and changing from a dim red dwarf like Wolf 359 into a cold, dead globe.

It all sounded beautifully simple. Unfortunately, serious complications have become evident, and it is now certain that the true life-history of a star is much more complicated than this. It seems definite, for instance, that the "power-house" deep inside the globe is a true power-house, and that the radiating energy of a star is not due solely to the heat set up by shrinking. The source of stellar energy is the rearrangement of the atoms which make up the body of the star.

What happens in the case of an ordinary main sequence star like the Sun is that nuclei of hydrogen atoms, which are far more plentiful than all the other types of atoms put together, build up into nuclei of another gas, helium. It takes four hydrogen nuclei to build one helium nucleus, and each time the combination occurs a certain amount of energy is let loose. It is this released energy which keeps the star radiating.

Of course, the actual process is extremely complex, and to enter fully into the mechanism would be beyond our present scope. Moreover, some of the giant stars have different ways of doing things, and even now our ideas are not at all clear-cut, since new theories are introduced almost every year—only to be rejected again when new facts come to light.

We are not even sure which are the youngest stars. Some of the intensely luminous Wolf-Rayet and B-stars are using up their resources so rapidly that they cannot have existed in their present condition for more than a few millions of years, whereas the comparatively leisurely Sun is much more economical, and has a life expectation of thousands of millions of years yet. It seems, however, that the Sun is not changing slowly into a red dwarf, as used to be supposed; as it ages, it is growing more luminous. After a blaze of glory, it may collapse, rather abruptly on the cosmical time-scale, into a small and incredibly dense star of the type known as a White Dwarf.

White Dwarfs are among the most curious objects in the

166

entire sky. They are certainly plentiful, but they are so faint that they are not easy to detect unless they are relatively close to us. Yet they are not unusually cool; some of them have peculiar spectra, indicating a surface temperature as great as that of Sirius, and much greater than that of the Sun. Their faintness must therefore mean that they are very small. One extreme example, Kuiper's Star, has a mass equal to the Sun but a diameter of only 4,000 miles, no more than that of Mars.

The mass of the Sun, but the diameter of Mars! There is only one way in which so much matter can be packed into so small a globe: the matter must be extremely dense. If a man could be taken to the surface of Kuiper's Star, he would find that he had a weight of 250,000 tons judged by our standards, while a thimbleful of the material of the star itself would weigh several thousands of millions of tons if it could be measured on the surface of the White Dwarf. Matter in such a state is completely beyond our understanding; we cannot conceive how it would be possible to pack so many tons weight into a thimble, and this amazing density has far-reaching results. For instance, the whole atmosphere of Kuiper's Star is probably less than twenty feet deep.

Near the centre of a normal star, the temperature is perhaps 14 million degrees, and the hydrogen-helium reaction can take place. But a White Dwarf has used up all its hydrogen, and it cannot shrink any further, so that it radiates feebly without any means of replenishing its energy. Whatever theory we adopt, it seems that the White Dwarfs are the old-age pensioners of the stellar system. They have even been described as "bankrupt stars".

At the other end of the scale, the vast, cool component of the double star Epsilon Aurigæ seems to have a diameter of some 1,800 million miles (Fig. 53). Here the mass is only 18 times that of the Sun, and the surface temperature a mere 1,200 degrees, while the inner heat is not great enough for the usual hydrogen-helium cycle to operate. The significant fact is, however, that the density must be as incredibly low as that of Kuiper's Star is incredibly high. It is impossible to draw Epsilon Aurigæ and a White Dwarf on the same scale; if we make the giant the size of this page, the dwarf will not be visible without a powerful microscope. Yet the masses will differ only by a factor of about 1 to 18.

In short, the stars are very unequal in size, in density and in luminosity, but not nearly so unequal in mass, as the large bodies are rarefied and the small ones dense. Few stars are known with 100 times the mass of the Sun, while the light-weights of the stellar system, the twins which make up the

o SUN

Fig. 53. Section of Epsilon Aurigæ (larger component) showing size compared with that of the Sun.

double star L726-8, still have $\frac{1}{25}$ of the solar mass, and are thus far more massive than Jupiter. Nature can play some strange tricks.

This shows, too, that there is a major difference between a small star and a large planet. Though Kuiper's Star is only the size of Mars, it is completely non-planetary in nature. Not only is it luminous, but it is far more massive than any planet could possibly be.

Planets moving round other stars are too faint to be observed directly, but are probably abundant. The first to be detected moves round our old friend 61 Cygni (or, more accurately, round one of the components of the 61 Cygni system); it seems to have a mass 15 times that of Jupiter, and was tracked down because it exerts a pull upon the visible star, affecting the star's proper motion. A few more cases are on record, but the most interesting is that of the nearby red dwarf, Barnard's Star. In 1963 P. van de Kamp was able to announce that Barnard's Star is associated with a body only $2\frac{1}{2}$ times as massive as Jupiter, and which must almost certainly be a planet.

It is impossible to do more than mention a few of the other curious bodies to be met with in the stellar heavens. Some stars seem to be surrounded by immensely distended atmospheres, while others, such as the remarkable object 48 Libræ, are "shell stars" with double atmospheres, the outer shell giving an impression of a flattened ring which puts one in mind of the ring-system of Saturn, though there is of course no real analogy. A few of the feebler Red Dwarfs appear to be subject to violent flares, so that they can increase perceptibly in brilliance and then fade back to normal over a period of only a few minutes. Then, too, there are the "supergiants" such as Canopus in Argo and Deneb in Cygnus, celestial searchlights with spectra that distinguish them at once from their milder fellows.

Recently much has been heard about "radio sources", which emit long-wave radiation and are studied by means of special apparatus. Oddly enough, these objects are not ordinary stars at all.

A radio telescope collects radio waves in roughly the same manner as an optical telescope collects light-waves ; no actual picture of the course is produced, but the information gathered is remarkably valuable, and radio astronomy has become one of the most vital branches of modern astronomical science—even though it began only in the 1930s (see Appendix XXVIII). Most people are familiar with the appearance of the great 250-foot "dish" at Jodrell Bank, but it is worth pointing out here that not all radio telescopes are built upon the dish pattern. Different investigations need different techniques and instruments.

Radio sources are of various kinds. The Sun, of course, is a powerful emitter of radio waves, and radiations have also been recorded from Jupiter (though, as we have seen, their exact nature is still uncertain). In our Galaxy, we have various objects such as the Crab Nebula in Taurus, which will be described below, and which is the wreck of a supernova—a star which suffered a cataclysmic outburst long ago. Most radio sources, however, lie far beyond our own star system, and are thus best deferred until Chapter Seventeen.

This book deals with optical astronomy; I am not a radio astronomer, and am not therefore competent to act as a guide to others. Yet it is worth noting that in radio astronomy, too, there is scope for the amateur, and in Britain there is one fully-fledged amateur observatory of this sort, built and used by F. W. Hyde at Clacton. The work carried out here has been widely recognized by professional scientists, and shows what may be accomplished by anyone with sufficient knowledge and practical skill.

In every field of research, then, there are real opportunities; and by learning something about what is going on in the heavens, we shall also form a better appreciation of the universe in which we live.

Chapter Fourteen

DOUBLE STARS

OF ALL THE the constellations in the sky, probably the best known is the Great Bear. It is not so brilliant as Orion, nor so spectacular as the Southern Cross; but it can always be seen in the northern United States when the night sky is clear, and most people have developed an affection for it. Besides, it is useful because two of its seven chief stars point to the Pole.

Even a casual glance will show something interesting about the "second star in the tail", known as Mizar or, on Bayer's system, Zeta Ursæ Majoris. Mizar itself is of the second magnitude, but close beside it is a much fainter star, Alcor, so dim that it is not particularly easy to see when there is the slightest haze.

Double stars of this kind are extremely common in the heavens, though most of them are too close together to be separated without the help of a telescope. They are spectacular enough to be well worth looking at for pure enjoyment, particularly when the two components are of different colours, and they are also useful for testing the performance of a telescope. A list of suitable "test pairs" is given in Appendix XXIII.

There are two classes of double stars. Sometimes the two members of a pair are not physically connected, so that the effect is due merely to the fact that one star happens to lie almost behind the other. One way to explain this is to picture two motor-cyclists coming down a long stretch of darkened road, using their headlamps and separated by perhaps half a mile. An observer watching them approach may well imagine that they are riding side by side, particularly if the nearer cyclist has the less powerful lamp. However, "optical" double stars of this type are not so common as might be imagined.

The physical connection between the components of some doubles was first realized a century and a half ago by Sir William Herschel. Actually, Herschel made the discovery more or less by chance. He was trying to measure the distances of some of the stars by the parallax method (see page 159), and he had made long series of observations of pairs which he thought might show an annual shift. He failed in his main object, because his instruments were not sufficiently accurate; but he did find that many of the doubles formed physically connected systems, and were in orbital motion round each other. Nowadays these genuine pairs are known as "binary stars".

It is not correct to say that the less massive star of a binary system revolves round its senior companion. Though the two may be very unequal in size and brilliance, they will certainly not be violently unequal in mass, since—as we have seen—the stars are strangely uniform in this respect; indeed, the smaller component may well be the more massive of the two. What will happen is that the two bodies will move round their common centre of gravity, much as the two bells of a dumb-bell move when twisted by their jointing arm.

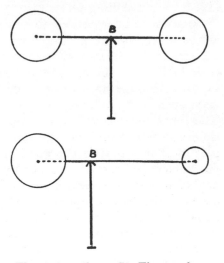

Figs. 54 and 55. In Fig 54 the bells are equal in mass, and the balancing point B is midway between them; in Fig. 55 the bells are unequal in mass, and B is no longer at the mid-point of the joining bar.

If the two components have equal mass, the centre of gravity will lie half-way between them, just as we can balance the dumb-bells by the middle of the arm (Fig. 54). If one star is the more massive, the centre of gravity will be displaced towards it (Fig. 55). The Earth and Moon move in this way; but since the Moon is so much the less massive, the centre of gravity of the system lies some way inside the Earth's globe.

A pair of binoculars will show that many apparently single stars consist of two, and a small telescope will reveal hundreds of pairs. Sometimes the components are equal, so that they are genuine twins, but more often one star is much brighter than the other. If a brilliant body is concerned, it may tend to drown its companion in a blaze of light, so that a telescope of some size will be needed to show both objects. Sirius is an excellent example of this. The brighter component is the most brilliant star in our skies, and it overpowers the White Dwarf companion, even though the White Dwarf would be an easy telescopic object were it shining on its own.

Binary stars have proved to be most useful to the theoretical astronomer. The orbits can be worked out; and as soon as the distance and the period of revolution are known, the combined mass of the stars in the system can be derived. Suppose, for instance, that the stars in a pair lie at an average distance from each other of 93 million miles, and have a period of one year. The Earth revolves round the Sun at this distance and in this time, and this means that the combined mass of the Sun-Earth pair must be equal to the combined mass of the two stars in the binary. In practice, we can neglect the Earth, which is of negligible mass when compared with any star, and in the above instance the two components of the binary would together equal one body the mass of the Sun. Unfortunately, calculating the separate masses of the components is not so straight-forward.

The whole method depends upon careful measurements of the apparent relative motions of the twin stars, and it is therefore not surprising that most of the bright pairs have been so closely studied by professional astronomers that there is not much point in the amateur's observing them further. Yet some of the generally-accepted measures are out of date, and there is definite scope for the serious observer with adequate equipment.

The separation of a double star is measured in seconds of arc. When it is borne in mind that the apparent diameter of the Moon is about half a degree, or 1,800 seconds of arc, it is evident that a pair of stars with a separation of only a second or two will need a powerful telescope if it is to be split. The apparent distance between Mizar and Alcor is roughly 700 seconds, but when a telescope is used the bright star is itself seen to be double, made up of two components between 14 and 15 seconds of arc apart. Actually, the system is more complicated even than this.

There is a minor mystery connected with Alcor. The old Arab astronomers called it "a test for keen eyes", but nowadays it can be seen by any normal person when the sky is clear and it can in no sense be regarded as a test. Either Alcor has brightened up during the last thousand years, or else it is not the star referred to by the Arabs. The real test star may be the much fainter object lying between Mizar and Alcor. This star is usually below the 8th magnitude, and thus quite invisible without a telescope, but it has been suspected of variability.

The "position angle" of a double star, binary or otherwise, is the direction of the fainter star as reckoned from the brighter, beginning with o degrees at the north point and reckoning round by east (90 degrees), south (180), and west (270) back to o, as shown in Fig. 56. This is generally enough to enable one to form a mental picture of the double before one actually goes to a telescope, though in the case of

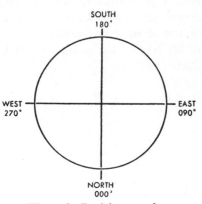

Fig. 56. Position angle.

perfect twins it is not easy to tell which of the components is meant to be the senior partner.

Measuring the separations and position angles of double stars cannot be undertaken with a telescope of less than 6 inches in aperture, and it is also necessary to have an equatorial mount, a driving clock, and a measuring device known as a micrometer. Micrometers are of various types; to describe them here would be beyond our scope, but full information can be found in the works listed in Appendix XXX.

The most beautiful of the double stars are those which show contrasting colours. Pride of place must go to Albireo or Beta Cygni, the faintest star in the "cross" of Cygnus, the Swan (Map VIII). The main star is of the third magnitude, and is of a strong golden-yellow colour, while the fifth-magnitude companion is a glorious blue-green. The two are sufficiently wide apart to be well seen in a 2-inch telescope, and a power of 50 on a 3-inch refractor will show them excellently. Other yellow and green pairs are known, but in my opinion, at least, none can rival Albireo.

There are also cases of bright orange-red stars, usually of Type M, which are accompanied by small green companions. Antares, leader of the Zodiacal constellation of Scorpio, is one of the reddest of the brilliant stars—its very name means "Rival of Mars"—and has also the distinction of being one of the largest giants known, so that in itself it is remarkable. Its beauty is enhanced by the fact that a small telescope will reveal an emerald-green star close beside it. The greenness of the faint companion is due partly to contrast with the ruddy hue of the giant, but it is none the less spectacular for that.

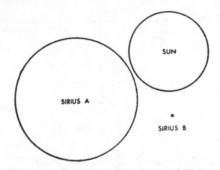

Fig. 57. Comparative sizes of Sirius A, Sirius B and the Sun.

Now and then we meet with some oddly-assorted pairs. One of the most interesting is Sirius, the Dog-Star (Map V). The main component is an A-type giant with a luminosity 26 times as great as the Sun's, and a diameter of more than a million miles. The second star could hardly be more dissimilar; it is a White Dwarf, considerably smaller than Uranus, but with a mass almost equal to that of the Sun. In Fig. 57, the sizes of the two companions are shown, with the Sun added for comparison.

Since Sirius has always been known as the Dog-Star, the White Dwarf companion has acquired the nickname of "the Pup", but at least it is a pup which can make its presence felt. Like Neptune, it was tracked down by its gravitational pull long before it was actually seen. Bessel, famous as the first astronomer to measure the distance of a star, found that Sirius itself was wobbling slightly in the heavens, and he calculated that this must be due to the effect of an unseen companion.

Years later, in 1862, the Pup was discovered, quite by chance, by an American instrument-maker who was testing a large new telescope.

Though the two stars of the Sirius pair are so unequal in size and luminosity, the bright giant has a mass only 2½ times as great as that of the White Dwarf. The distance between the two is about equal to that between Uranus and the Sun, and the period is about fifty years, so that two complete revolutions have been completed since the Pup was first seen. As a matter of fact, the Pup does not appear to be particularly faint, but it is not easy to observe, since the glare from the larger star drowns it. It has been claimed that a 6-inch telescope will show it, but I admit that I have yet to see it with my 12½-inch reflector, probably because Sirius lies well south of the celestial equator and so never rises high above the horizon in England. I have, however, seen it with the 24-inch reflector made and used by G. A. Hole in Sussex.

Some double stars are too close to be split with any telescope, but can nevertheless be detected by means of our old and reliable ally, the Doppler Effect. In the very much over-simplified diagram given in Fig. 58, it is assumed that the fainter star (B) is revolving round the brighter (A). In position 1, B is moving towards us, and its spectrum will show a violet

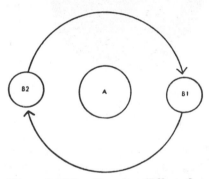

Fig. 58. The Doppler Effect for spectroscopic binaries.

shift; in position 2, it will be receding, and the shift will be towards the red. Consequently, the combined spectrum due to the two stars will show variations, and the binary nature of the system will be betrayed. Even if the spectrum of one component is too faint to be seen at all, the wobbling of the lines of the other star will be just as tell-tale. Pairs of this kind are termed "spectroscopic binaries".

Now and then we meet with positive family parties of stars,

systems including three, four or even six components. One of the best known is Epsilon Lyræ, shown in Map VIII, lying close to the brilliant star Vega, which appears almost directly overhead in England during summer evenings. Keen eyes can see that Epsilon is made up of two components, and in binoculars the pair can be well seen, since the apparent distance between them is 207 seconds of arc. A 3-inch telescope reveals that each component is again double, so that there are four visible stars in the system (Fig. 59), and to make things even more complex one of the four is itself a spectroscopic binary. The two main pairs are so far apart that they take at least a million years to complete one revolution around their centre of gravity.

Fig. 59. The famous "double-double" star Epsilon Lyræ.

Equally remarkable is Castor, one of the main stars in the famous constellation of Gemini, the Twins (Map V). Here we have two bright components at present 1·8 seconds of arc apart, though the revolution period is 380 years and it is not now so easy to split the pair as it used to be half a century ago. Each is a spectroscopic binary, and there is a 9th-magnitude spectroscopic binary companion 73 seconds of arc away, so that the system of Castor is made up of six separate suns.

On the other hand, Gamma Virginis, in the Y of Virgo (Map VI) is at present a grand, easy double separable in a very small telescope. By the end of the century it will have closed up so much that it will appear single in ordinary instruments.

The magnification for looking at any particular double star must depend upon the individual double itself. If you want to obtain an overall view of Mizar and its companions, a low power is necessary, since if you increase the magnification you will find that Alcor is out of the field. Closer pairs naturally need higher powers, and for measuring work considerable magnification must be used.

Useful research can be carried out by the amateur double-star observer; there is still routine work to be done, and in any case there is much enjoyment to be gained from looking at the pairs and groups of suns. With their varied separations and their lovely contrasting colours, they are among the most beautiful of the objects in the stellar heavens.

Chapter Fifteen

VARIABLE STARS

FORTUNATELY FOR US, our Sun is a steady, well-behaved star. It may have periods of unusual activity, when its disk is disturbed by spot-groups and flares, but at least its output of energy does not alter greatly over the lapse of hundreds of centuries.

Other suns are not so quiescent. Some of them vary in brightness from day to day, even from hour to hour, either regularly or in an erratic manner. They swell and shrink, and their temperatures change with their fluctuations, so that any planet circling round them would be subject to most uncomfortable changes of climate.

Variable stars are important both to the professional and to the amateur, and the owner of a small instrument can do useful work, particularly as his telescope need not be so perfect as that of the lunar or planetary observer (though, of course, the better the telescope the better the results). It is true that the regular variables of short period have been closely studied at the great observatories, but there are other stars which seem to delight in springing surprises, so that they need constant watching.

It is not easy to give a general classification of the different types of variable stars. However, the following rough notes may be useful as a guide.

First there are the eclipsing binaries, such as Algol in Perseus, which are not true "variables" at all, even though they do seem to alter in brightness. Perhaps the most important of the true short-period variables are the Cepheids, so named because the star Delta Cephei is the best-known member of the class; the periods range from a few days up to six or seven weeks. Of much shorter period are the RR Lyræ stars, whose periods range between 30 hours and less than 2 hours. Then there are the long-period variables, usually Red Giants of great size and comparatively low temperature, with periods ranging from 70 days to over 2 years. Irregular variables, as their name suggests, behave in an unpredictable manner. Lastly come the violently explosive "temporary stars" or novæ.

There are several variables which can be followed without any telescope at all. The most famous of these is Betelgeux, the Red Giant in Orion. It belongs to the irregular class, though there is a very rough period of from 4 to 5 years, and it changes in brightness from magnitude 0 down to 1·2,* so that whereas it may sometimes outshine the glittering white Rigel it may at others become fainter than Aldebaran, the "Eye of the Bull". The alterations are slow, but they become noticeable over a week or two, and the beginner who estimates the magnitude of Betelgeux every few days will soon be able to detect the fluctuations. However, most of the interesting variables cannot be followed without a telescope, since when near minimum they are below naked-eye visibility.

Before coming to the proper variables, it will be of interest to say something about the "fake variables", or eclipsing binaries. These might well have been described in the chapter dealing with double stars, but since they do seem to change in brilliancy they come under the scope of the variable star enthusiast.

The best-known of these "fakes" is Algol, which lies in the constellation of Perseus and is shown in Map VII. In mythology, Perseus was the hero who slew the fearful Gorgon, Medusa, whose glance turned the hardiest onlooker to stone,† and it is fitting that Algol should mark the Gorgon's severed head.

Usually Algol shines as a star of magnitude 2·1, only a little inferior to Polaris. It remains constant (or virtually so) for a period of 2½ days, but then it starts to fade, until after about five hours it has dropped to magnitude 3·3. After a relatively brief minimum, it starts to brighten once more, taking a further five hours to regain its lost lustre. Textbooks usually say that its variations were discovered by Montanari in 1667, but the old Arab astronomers called Algol "The Winking Demon", which is interesting if they were unaware of its odd behaviour—as they seem to have been.

* In a recent catalogue of variable stars, that of Kukarkin, the greatest brilliancy of Betelgeux is given as magnitude 0·4; but past records seem to show that on rare occasions the star can attain magnitude 0·1 or even 0·0, brighter than any other stars in the sky except Sirius, Canopus, Alpha Centauri and Arcturus.

† Nowadays, this power is possessed only by the Chancellor of the Exchequer.

Algol is not truly variable. The apparent fluctuations are due to the fact that the system is a binary, and when the brighter star is eclipsed by the fainter the total brightness naturally drops. When the fainter star is obscured by the brighter, there is a small minimum, but since this amounts to only one-twentieth of a magnitude it cannot be detected with the naked eye. Actually the system of Algol includes a third star, but the principle of the variations is straightforward enough.

The beginner may like to plot Algol's "light-curve". A light-curve is merely a graph plotting time against magnitude, as shown in Fig. 60, and it is always interesting to make one from personal observations. In the diagram of Algol given here, the secondary minimum is slightly exaggerated, as otherwise it would not be visible upon a chart drawn to so small a scale.

Another bright eclipsing binary is Beta Lyræ, which lies near the brilliant Vega (Map VIII). Here there are two bright components, so close together that they almost touch, and in consequence too close to be seen separately in any telescope. At maximum, when both stars are shining together, Beta Lyræ appears of magnitude 3·4. It then fades steadily to magnitude 3·8, and then rises once more to 3·4, but at the next minimum it descends to 4·4, so that deep and shallow minima take place alternatively. The brightness is always varying, so that there is no long comparatively steady maximum, as with Algol. One remarkable fact about the components of Beta Lyræ is that each is stretched out into the shape of an egg, simply because the two stars are pulling so strongly on each other; the general situation has been compared to two eggs rolling about with their sharper ends kept close together.

Epsilon Aurigæ, the giant of giants, is also an eclipsing binary (Map IV), and in fact the larger component is the fainter of the two. The period is over 27 years, the longest known for any eclipsing star. Its neighbour in the sky, Zeta Aurigæ, has a period of 972 days, and is particularly interesting to spectroscopic workers because the smaller star shines for some time through the outer layers of the diffuse giant component before disappearing behind. The fluctuations of Epsilon and Zeta Aurigæ are however much less obvious than those of Algol, and are not marked enough to be noticeable with the naked eye.

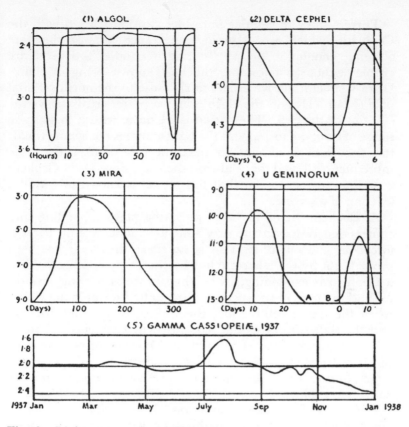

Fig. 60. Light-curves of variable stars.
  (1) The eclipsing binary Algol.
  (2) Delta Cephei, the prototype Cepheid.
  (3) Mira Ceti. As explained in the text, the period and amplitude
      of variation are not constant.
  (4) U Geminorum. A semi-regular variable. Maxima are of two
      types, "short" and "long". The distance AB in the graph
      represents a time interval which may be as little as 40 or as
      long as 130 days.
  (5) Gamma Cassiopeiæ, 1937. These estimations were made by
      the writer, using the naked eye only. It is unlikely that they
      are at all precise; however, they suffice to show that obvious
      fluctuations were going on. The star is irregular, and is
      classed by Kukarkin as a "pseudo-nova".

Turning now to genuine variables, we must begin with the Cepheids, which are of great importance because they are obliging enough to act as "standard candles". Several are visible without a telescope, the best known being of course Delta Cephei itself, which lies fairly close to the north celestial pole (Map VII) and therefore remains permanently above the horizon in the U.S. The period is 5⅓ days, with a magnitude range of from 3·5 to 4·4, and the light-curve is not symmetrical; the rise from minimum to maximum is quicker than the subsequent fall, and this is always the case, since Delta Cephei's variations are so regular that the period is known to within a fraction of a second.

A Cepheid seems to be a pulsating star, expanding and contracting rather in the way that a balloon will do if air is forced in and out of it. This is no mere theory; it has been proved, not by the telescope but by the Doppler Effect. When a Cepheid is expanding, its bright surface is moving towards us, and the lines in the spectrum are shifted towards the violet; when the star is contracting, the surface is receding from us, and the shift is towards the red. Just why Cepheids pulsate in this way is something of a mystery. Many theories have been put forward, but to describe them in detail would take many pages, and in any case none of the present theories seem to be entirely complete.

Equally mysterious is the Period-Luminosity Law, which has provided the stellar astronomer with one of his most powerful weapons. Reduced to its simplest terms, this Law links the variation period of a Cepheid with the star's actual luminosity, so that variables of equal period have the same candle-power. Delta Cephei, period 5⅓ days, is approximately 660 times as luminous as the Sun; therefore, every Cepheid with a period of 5⅓ days is 660 times as luminous as the Sun.

This by itself would be intriguing enough, but it has far-reaching consequences. If we know the real brightness of a distant lighthouse, and we can measure how bright it appears to be, we can work out its distance from us by means of simple arithmetic. In the case of Delta Cephei, we know its real luminosity and its apparent magnitude, so that its distance follows at once; it proves to be 1300 light-years. In fact, we can find out the distance of any Cepheid merely by watching how long it takes to vary from maximum to maximum.

Though we do not know precisely why the Law operates, there is no doubt of its validity. The longer the period, the more luminous the star. These strange variables are the standard candles of the universe, and they never depart from their own rules. There is no such thing as an unconventional Cepheid.

Consequently, we find that we have the means of measuring the distance of a remote star-cluster or galaxy. If we can detect a Cepheid, we can find its distance, and so the distance of the cluster which contains it must be the same. Nature can be awkward at times, as we know to our cost, but in this case she has given us an unexpectedly accurate measuring-rod. There are certain complications, since there seem to be two different types of Cepheids with rather different period-luminosity relationships, but refinements of this nature do not seriously mar our space-gauging.

RR Lyræ stars were formerly classed with the Cepheids, but it now appears that they form a separate group. Their fluctuations are perfectly regular, and their periods range from only 1½ hours up to slightly more than a day. All RR Lyræ stars have about the same luminosity, roughly 85 times that of the Sun, so that they too can be used as standard candles. All are distant, and so appear too faint to be easily studied by the amateur observer.

The periods of the eclipsing binaries, the Cepheids and the RR Lyræ stars, are known so accurately that there is no point in the amateur's observing them further. Nor do any reasonably bright variables of such types remain to be discovered. On the other hand, the long-period stars present very different problems. They are not perfectly regular, and they are not so closely studied by professional astronomers, so that here the amateur can come into his own.

In August 1596, David Fabricius recorded a third-magnitude star in the constellation of Cetus the Whale, not far from Orion (Map IV). By October it had disappeared. Bayer saw it again in 1603, when he was drawing up his star catalogue, and gave it the Greek letter Omicron, but shortly afterwards it vanished once more. Not until some time later was it found that the star appears with fair regularity; it takes almost a year to pass from maximum to maximum, and it is visible to the naked eye for many weeks at a time. Not unnaturally, it was given the name of Mira, "The Wonderful".

The period of naked-eye visibility is not always the same, and nor is the magnitude at maximum. At some maxima the star attains the second magnitude, and remains visible without a telescope for over 20 weeks, but in other years it becomes no brighter than magnitude 5. In 1868, for instance, it was a naked-eye object for only 12 weeks. Near minimum the magnitude falls to below 9, so that Mira cannot then be found even with binoculars or a small telescope. Nor is the period constant; the 331 days given in most textbooks is merely an average, and may fluctuate to the extent of more than a month either way. There is nothing neat or precise about "the Wonderful Star", and for this reason alone it is worth keeping under watch. Extra interest is added by the tiny white companion, so faint that it is hard to see except when the senior star is near minimum.

Like all long-period variables, Mira is a Red Giant, large, cool and diffuse. Many similar stars are known, some of which can be seen with the naked eye when at their brightest, and our knowledge of their behaviour depends mainly upon the results of amateur work. There is no period-luminosity law, and thus the stars cannot be used as standard beacons, with the result that professional astronomers do not study them so closely as in the case of the Cepheids. Again there is a pulsation as well as a change in temperature, but the underlying cause of this pulsation is not known. It is rather strange to find that stars with the longer periods often prove to be of relatively low luminosity.

Though the long-period stars are at least partly regular, there are some variables which seem to be completely erratic. These irregular variables are perhaps the most fascinating of all, since one never knows what they are going to do next. Betelgeux is one example, and other Red Giants which behave in a similar way are Alpha Herculis (Map IX) and Mu Cephei (Map VII). Mu Cephei is particularly interesting. It varies between magnitudes 3·6 and 5·1, so that it can always be seen with the naked eye, but a pair of binoculars will show that it is of a beautiful red colour. It looks almost like a drop of blood, and it deserves the name of "the Garnet Star" given to it by Sir William Herschel.

Cassiopeia, the Queen, is one of the most prominent of the northern constellations, and few people can mistake its five chief stars, which are arranged in the form of a rough W (Map VII). The middle star of the W, Gamma Cassiopeiæ, is an interesting variable. It used to be ranked as a steady body of magnitude 2·3, but in 1936 it abruptly brightened up by over half a magnitude, so that it far outshone Polaris. Since then it has varied between magnitudes 2 and 3·3. Its spectrum is so peculiar that it cannot be placed in any ordinary type.

Telescopic irregular variables are of various classes. Stars like RV Tauri have alternate bright and fainter maxima; U Geminorum stars remain at minimum for long spells, with abrupt and unexpected rises to short-lived maxima, whereas stars of the R Coronæ Borealis class remain at maximum for most of the time, with occasional minima. Each irregular variable has its own characteristics, as the serious observer will soon find out.

No description of variable stars would be complete without some reference to the madcap of the heavens, Eta Argûs.* It lies in the southern constellation of Argo Navis, the Ship, and never rises above the English horizon, which is unfortunate. For the last 80 years it has been just below naked-eye visibility, but for a while, between the years 1837 and 1854, it ranked among the most brilliant stars in the sky, and was surpassed only by Sirius and Canopus. It is in a class by itself, and it too has an unusual spectrum. Eta Argûs lies in a cloud of gas, and the erratic variations may be due partly to the differing depth of the gas-layer between us and it; but the star must vary intrinsically as well, so that it may at any time jump back into prominence (see Map XIV).

Variable star observations are made by estimating the magnitude of the variable as compared with near-by stars of known brightness. For instance, Gamma Cassiopeiæ is provided with two perfect comparison stars in the same constellation, Beta (magnitude 2·26) and Delta (2·67). In the case of a telescopic object, the comparison stars must of course lie in the

* Since the great constellation of Argo has been divided up, Eta Argûs has been re-christened "Eta Carinæ", while Canopus has become "Alpha Carinæ".

same field as the variable, and a few awkward stars which lie aloof by themselves are not easy to estimate properly.

The first thing to do is to identify the variable. A star atlas is necessary, Norton's being by far the best for the ordinary observer, and the position of the variable can be plotted (if, of course, it is not already shown). It is however a mistake to look directly for the variable itself. The best method is to note the stars which will be found in the same low-power field, so that an overall impression can be built up. Most long-period variables stand out because of their redness, but this is never a safe guide, and is in any case not valid for the short-period stars and the irregulars.

It may sound difficult to identify any particular starfield, but no two fields are alike, and a little practice will work wonders. It is sometimes suggested that the best way is by "sweeping about" until the required field comes into view, but this is a mistake. When a telescopic variable is to be sought, there should be a definite plan of campaign. First identify the area by means of naked-eye stars which can be recognized without possibility of error, and then proceed by means of star alignments and patterns, swinging the telescope north and south and in right ascension in terms of a known angular field. In difficult cases, an easily recognizable star can be selected which has the same declination as the variable, and the telescope left stationary until the variable drifts into view (though slight adjustments will be needed if the telescope is mounted on an altazimuth stand). It is unwise to leave any "safe anchorage" for the next until it has been identified with absolute certainty. When an observer has once found the field, he will usually recognize it again without much trouble, and it can be picked up in a matter of seconds, but the approach should always be "planned". A moment's carelessness can lead to some very peculiar results.

If the observer belongs to an astronomical society, he can of course obtain charts of the fields he needs. Approximate positions of some of the long-period and irregular variables are shown in the star maps given on pages 264–308.

There are several methods of making estimations. One of the simplest is Pogson's Step Method, in which the observer trains himself to gauge a difference of 0·1 magnitude, which constitutes one "step". Suppose that he is observing a variable star, and finds that it is two steps fainter than comparison star A and one step brighter than comparison star B. He records: "A —2: B + 1." He then looks up the magnitudes listed for A and B. If A is 8·0 and B 8·3, the variable must be 8·2, which is two-tenths of a magnitude fainter than 8·0 and one-tenth of a magnitude brighter than 8·3.

A more complex method is the Fractional, used by many workers. Here two comparison stars are used, and the brightness difference between them is divided mentally into a convenient number of parts, after which the variable is placed in its correct position in the step-series. If A is the brighter of two comparison stars A and B, and the variable is estimated as one-quarter of the way from A to B (and hence three-quarters of the way from B to A), the record will read: A 1 V 3 B. The magnitudes of the comparison stars can then be looked up as before, and the magnitude of the variable worked out.

There are many points to bear in mind when using either of these methods, and perhaps the most important is that the observer should go to his telescope with an open mind. If he expects the variable to be of magnitude 7·5, there is a strong chance that he will in fact record it as 7·5, whether this is correct or not! Neither is it easy to compare a red star with a white one. Plenty of practice is needed, but the serious enthusiast will soon find that he has "got the hang of it", after which he will be able to estimate many variables during the course of a few hours' work.

One difficulty of observing naked-eye variables such as Betelgeux is that a star is bound to be reduced in brightness as it approaches the horizon, since it will be shining through a thicker layer of the Earth's atmosphere. This "extinction" effect can upset an observation completely if it is not allowed for, but the table given in Appendix XXIV should help. With telescopic observations, extinction can be neglected, since all the stars in the field will be at approximately the same altitude above the horizon.

The so-called "secular variables" are of very different type. They are stars which seem to have undergone a slow brightening or fading over the course of centuries. For instance, the faintest of the seven stars in the Great Bear, Megrez, used to be as bright as its companions, and was so recorded by Ptolemy; but it is now a magnitude fainter, though it has been suspected of slight fluctuations and is thus well worth watching. Castor, one of the famous Twins, is now fainter than Pollux, though it used to be brighter; and Theta Eridani, in the River (Map XI), has sunk from the first magnitude to the third since the *Almagest* was drawn up. On the other hand Alhena or Gamma Geminorum, not far from Castor (Map V), has brightened up from the third magnitude to above the second. Alterations in these stars are too slow to be detectable during the course of a life-time, and in any case we cannot place full trust in the old estimates, but there is always a chance of observing something unexpected.

If the secular variables are leisurely, the "temporary stars" or Novæ are nothing of the sort, and are perhaps the most violent objects in the entire universe. Occasionally a star will blaze up where no star was seen before; it may attain great brilliance for a few days or a few weeks, but eventually it will fade back into insignificance, becoming so dim that it will be hard to see even with a powerful telescope.

It used to be thought that a nova was really a new star, as the name suggests, but this is a mistake. What happens is that a normally faint star undergoes a tremendous outburst that results in a 70,000- or 80,000-fold increase in brightness. One theory held that the flare-up was the result of a collision between two stars, but novæ are not uncommon (even though few of them reach naked-eye visibility), and the stars are so widely scattered in space that the idea of frequent stellar collisions is quite ruled out. More probably a disturbance inside the star causes the outer layers to be blown off with explosive violence, and this idea is supported by the fact that some novæ develop gaseous surrounds of incıedibly large size and correspondingly low density. When we come to ask the reason for the explosion, however, we have—as usual—to confess that we are uncertain. If our Sun became a nova, the results from our point of view would be decidedly unfortunate, but luckily the Sun seems to be refreshingly stable.

Nineteen naked-eye novæ and many fainter ones have been seen during the present century. Pride of place must go to Nova Persei 1901 and Nova Aquilæ 1918, each of which became brighter than any stars in the sky apart from Sirius and

Canopus, but which have by now become very faint telescopic objects. More recently we have had Nova Herculis 1934, which, like many other novæ, was discovered by an amateur observer. It was found on December 13 by J. P. M. Prentice, then Director of the Meteor Section of the British Astronomical Association, who had been observing shooting-stars and was taking a nocturnal stroll after finishing his regular programme. The star had an unusually long maximum, and as it faded it developed a strong greenish hue, which was most striking with the 3-inch refractor that I was using at the time.

Probably the most interesting nova of modern times was found by G. E. D. Alcock, a British amateur, in the constellation of Delphinus in July 1967. The discovery was not purely accidental. Using binoculars, Alcock had been making systematic searches for novae and for comets; he had already been rewarded with four comets, but Nova Delphini was even more important. It was of about the fifth magnitude, and therefore dimly visible to the naked eye, but instead of fading quickly, in the manner of most novae, it remained near maximum brightness for months.

Discoveries of this sort cannot be made by sheer chance. A bright naked-eye nova forces itself on one's attention; Nova Delphini did not. Alcock had spent years learning his way around the sky, and he can identify some 30,000 stars visible with his binoculars. His achievement was a particularly good example of the way in which the dedicated amateur observer can make real contributions to science.

Normal novæ are spectacular enough, but the rarer "supernovæ" are even more so. Here the increase in light is much greater, and at maximum a supernova may shine as brightly as all the other stars in its system put together. In our own galaxy, the most famous supernova on record is that of 1572. It lay in Cassiopeia, and at its brightest was more brilliant than Venus, so that it remained visible in broad daylight. Telescopes still lay in the future, so that as soon as the star fell below the sixth magnitude it was lost to sight, and it cannot now be identified with certainty. However, a certain amount of "radio emission" from the area may mark the place where the supernova once blazed. Another supernova, that of 1054, has left a visible cloud of gas which is now called the Crab Nebula, and this too is a powerful radio emitter. The only other super-novæ to be seen in our own system during recorded times were those of 1006 and 1604.

Normal bright novæ appear only at intervals of years, as is shown in the list given in Appendix XXV, but there is no harm in the amateur's occupying himself for four or five minutes a night in making a naked-eye survey of the Milky Way zone. Probably he will never make a startling discovery, but he will at least improve his knowledge of the sky, and there is always the remote chance that he will achieve lasting fame.

I have already mentioned the red dwarf "flare stars", which may brighten up appreciably over a period of a few minutes, and fade back to their normal brilliance within the hour. Recently they have been the subject of much attention by both optical and radio astronomers, and the visual ob-servers at the Crimean Astrophysical Observatory have been co-operating with the radio astronomers at Jodrell Bank; similar work has been carried out in the U.S.A. The trouble here is that unless flare stars are kept under continuous watch, their outbursts will be missed. All are faint, and generally speaking are best studied with photoelectric equipment combined with large telescopes. However, it may well be that amateurs will be able to make a useful contribution to flare-star studies, always provided that adequate instruments are avail-able—together with an almost inexhaustible store of patience!

It is clear that variable stars can give the observer plenty to do. There are the red stars of long period, the irregulars with their quirks and eccentricities, and the occasional strange novæ which flare up to unexpected brilliance. The stellar heavens are never dull, and there is always something new to see.

Chapter Sixteen

STAR-CLUSTERS AND NEBULÆ

SOME WAY FROM Orion, beyond the bright red star Aldebaran, can be seen what at first sight looks like a faint misty patch. Close inspection shows that this patch is in fact made up of stars, one of which is of the third magnitude and the rest much dimmer.

Seven stars can be made out by normal-sighted people, and the group is known popularly as the "Seven Sisters",* though its official name is the Pleiades. It is a genuine cluster, and not a line-of-sight effect. It has been calculated that the odds against any chance alignment of the seven most conspicuous stars are millions to one against (see Plate XII).

The Pleiades have been known from very early times, and legends about them are found in ancient mythology, but it is only during the last three and a half centuries that astronomers have realized that there are many similar clusters in the sky. One or two can be seen without optical aid; there are the Hyades round Aldebaran, Præsepe or the "Beehive" in Cancer, and the Sword-Handle in Perseus. Most, however, are too faint to be seen without a telescope.

A pair of binoculars will show the Pleiades very well. With a magnification of about 20, the chief stars fill the field, and look like jewels gleaming against black velvet. Moreover, fainter stars jump into visibility; even a small telescope reveals so many that to count them would be a difficult process. The Seven Sisters have many junior relatives—over 250, in fact.

The Pleiad stars look close together, but the cluster is not really so dense as might be imagined, though if the Sun lay in the middle of the Pleiades our sky would contain many stars shining more brightly than Sirius does to us. Nor must we be deceived by the fact that the whole cluster takes up only a small patch of the heavens, since the real diameter of the group is over 15 light-years. Its distance is 490 light-years.

* Really keen-sighted people can see up to a dozen Pleiads without optical aid, but artificial lights make it difficult to see the cluster as anything but a dim glimmer.

Almost as famous as the Pleiades are the Hyades, which lie around Aldebaran itself, and are shown in Map IV. Actually, Aldebaran is not a genuine member of the cluster, as it merely happens to lie in the same direction. Telescopically the Hyades are not so beautiful as the Seven Sisters, as the stars are much wider apart, and it is difficult to get them all into the same field of view. Moreover, they are overpowered by the bright orange-red light of Aldebaran.

There is an important difference between the two clusters. In the Pleiades, the brightest stars are blue, highly luminous giants of type B, whereas in the Hyades the chief members of the group are orange giants of type K. B-stars do occur in the Hyades, but are much less in evidence.

Another naked-eye "open cluster" is Præsepe, shown in Map V. It lies in Cancer the Crab, and has been nicknamed the Beehive, because in a small telescope it has been said to give some impression of a collection of luminous bees. It is not prominent without a telescope, and even a half-moon is enough to drown it, but it is a fine sight in a small instrument.

Even more striking are the twin clusters in Perseus (Map VII), marking the "sword-handle" of the legendary hero. To the naked eye the only indication of their presence is an ill-defined misty patch, but a telescope reveals two rich star-clusters in the same low-power field. I have found that a good view is obtained with a power of about 30 on a 3-inch refractor.

Telescopic clusters are numerous, and anyone equipped with a small instrument can give himself many hours of enjoyment by sweeping for them and learning how to pick them up. Each has a separate designation, most of the brightest being known by their numbers in Messier's catalogue. Charles Messier, it will be remembered, was the French comet-hunter who was constantly annoyed by confusing clusters with comets, and so drew up a list of objects to be avoided during his searches. Thus Præsepe is M.44, the Pleiades M.45, and the nebula in Orion M.42. The full catalogue, given in Appendix XXVI, contains 107 objects. A few of the objects listed by Messier cannot now be found, and may have been comets that the French observer failed to recognize for what they were.

Præsepe and the Pleiades are open clusters, but some of the objects recorded by Messier are of different type. There are for instance the globular clusters, which look like compact balls of stars, so closely crowded towards their centres that it is difficult to distinguish the individual points of light. A rich globular may contain a hundred thousand separate stars, and the crowding is much greater than in the case of the open clusters.

All globulars are very remote, and even the nearest of them lies at a distance of thousands of light-years. They form a sort of "outer surround" to our stellar system, and since the Sun is not in the middle of the Galaxy we naturally have a better view of the globulars to one side of the sky. Most of them appear round the southern constellations of Scorpio and Sagittarius.

The best way to demonstrate this effect is to imagine that we are standing in a woodland glade on a foggy evening. If we stand away from the centre of the glade, we can see the bordering trees to one side of us, but the trees which mark the edge of the glade on the far side will be concealed by the fog. If we take each tree to represent a globular, we can understand why these strange clusters are best seen in one particular direction. Space, too, is "foggy"; there is a good deal of interstellar dust and gas, and light-waves cannot penetrate it, so that in certain directions our view is blocked.

The globulars are too far away to have their distances measured directly, but fortunately they contain RR Lyræ stars, and these useful beacons give us the answers at once. As has been shown, we can find the distance of an RR Lyræ star simply by watching how long it takes to pass from maximum to maximum, and the distance of the globular in which it lies naturally follows. Originally there was some confusion because RR Lyræ variables were thought to follow the Cepheid period-luminosity law instead of having one of their own, but this misunderstanding has now been cleared up.

The brightest of the globular clusters visible in the U.S. is M.13, situated in Hercules (Map IX), which is faintly visible to the naked eye on a clear night. Binoculars will show it as a hazy patch, and a 3-inch will reveal stars near its edges, but to see it really well one needs an aperture of from 8 to 12 inches. Then, even the centre can be seen to consist of a myriad tiny points, and the sight is superb. Oddly enough, M.13 is unusually poor in RR Lyræ variables.

Globulars are much less common than the open clusters.

Only about 100 are known, and most of these are faint, so that Messier listed only 28 of them. Unfortunately for us, the two finest globulars, Omega Centauri and 47 Tucanæ, lie too far south to be visible in the U.S.

Messier was not really interested in clusters, and regarded them simply as nuisances from the cometary point of view, so that he listed only those which were liable to confuse him. Since his day, many more of the hazy patches have been catalogued, until by now the total number runs into millions. Herschel discovered many between 1775 and 1820, and he saw that they were of different kinds. Some were obvious clusters, but others looked more like filmy gas, and these latter were termed "nebulæ", from the Latin word for "clouds".

For many years, it was believed that the nebulæ were merely star-clusters so far away that they could not be resolved with the telescopes available. This also applied to the curious "planetary nebulæ", so called because they showed pale disks not unlike those of the planets. But doubts began to creep in; some of the nebulæ did not look in the least like clusters, and their real nature remained dubious.

By itself, the telescope could not solve the mystery, but the spectroscope came to the rescue. In 1864 Sir William Huggins, one of the great spectroscopic pioneers, put the matter to the test by observing a planetary nebula in Draco. He half expected to see a somewhat confused effect due to the result of the combined spectra of thousands of stars, but instead he saw nothing but a single green line. At once he realized the truth. The light of the nebula was made up of one colour only, emitted by a luminous gas; the object was not a distant cluster at all, but something quite different.

Diffuse nebulæ such as that in Orion's Sword had always been regarded as clouds of gas and dust in space; the planetaries too were found to be gaseous, but they are neither planets nor nebulæ, so that their popular name could hardly be less apt. A typical planetary consists of a very faint, very hot Wolf-Rayet star associated with an immense "atmosphere" made up of incredibly tenuous gas. The low density is not easy to appreciate; if we took a cubic inch of air and spread it out over a cubic mile, we would arrive at about the correct value.

The brightest of the planetaries is the Ring Nebula, M.57

Lyræ, close to Vega (Map VIII). It is easy to identify, since it lies between two fairly bright stars, the famous eclipsing binary Beta Lyræ and its neighbour Gamma. A 3-inch telescope will show it, but a larger aperture proves that it has the shape of a ring, not unlike a faintly luminous motor-car tyre, while the central star is only of the fifteenth magnitude. Some of the other planetaries are much less symmetrical.

Planetaries are most interesting objects, and it is a pity that most of them are so dim. However, there are plenty of diffuse nebulæ. A few of them, notably M.42—the Sword of Orion—can be seen with the naked eye. M.42 lies below the three stars of the Belt, as shown in Map IV, and cannot possibly be missed. It is one of the show-pieces of the sky, particularly as it contains the celebrated multiple star Theta Orionis, known commonly as the Trapezium because of the arrangement of its four brightest components. M.42 itself is 15 light-years across, and about 1,800 light-years away.

Many of these diffuse nebulæ are within the range of a small telescope, but not all are of the same type. Some, such as the nebula contained in the Pleiades cluster, shine simply by reflecting the light of the intermingled stars, but others, including M.42 Orionis, show spectra which indicate that they are shining by themselves; the radiation from the mixed-in stars is affecting the gas and making it luminous. Like the planetaries, the diffuse nebulæ are very rarefied, millions of times less dense than our own atmosphere.

Spectroscopic work has led to our identifying many of the gases in nebulæ. Hydrogen, helium and oxygen are all present, and there are also spectrum lines which were once thought to be due to a new element "nebulium", but which have disappointingly proved to be due merely to oxygen and nitrogen in unfamiliar states.

Diffuse nebulæ shine because of the stars contained in them, and consequently a nebula that includes no convenient star will remain dark. Though it will therefore be invisible, it will make itself evident because it will blot out the stars or luminous gas behind it. Herschel was inclined to believe that the occasional well-defined starless patches were true "holes in the heavens", but it is now known that there is no basic difference between a nebula which shines and one which does not. Of the dark objects, the most prominent are the Coal Sack in the Southern Cross, unfortunately invisible in N. America, and a section of the Orion Nebula known as the "Horse's-Head" because its shape gives the impression of the head of a knight in chess. There is also a dark patch on Cygnus, not far from Deneb; and several others can be found with small apertures, though they are not striking.

An object that cannot be classed with either the planetaries or the diffuse nebulæ is M.1, the "Crab Nebula" in Taurus (Map IV). This too is bright enough to be seen with a small telescope, but a large instrument is needed to show it well. The gases in it appear to be expanding from a central point, and there is no doubt that the nebula is the wreck of the supernova that flared up in the year 1054. It is one of the most powerful known emitters of radio waves, and is also a source of X-rays, so that it is unusually interesting.

In general, a lower power is to be preferred for observing nebulæ, except for the planetaries; a magnification of 30 to 40 on a 3-inch refractor is quite high enough for most purposes. It is true that no useful work can be done, but this is no grave disadvantage. It is always worth while to relax and enjoy oneself among the wonders of the sky.

Chapter Seventeen

THE GALAXIES OF SPACE

ONE OF THE glories of the night sky is the luminous band which is known to everyone as the Milky Way. It stretches right round the heavens, and on a clear moonless night it is a magnificent spectacle.

Galileo's first telescope, applied to the sky in the winter of 1609-10, led him to say that the Milky Way is made up of "an infinite number" of stars. This is an exaggeration; the stars are not infinitely numerous, but there are about one hundred thousand million of them in our own system, together with a vast quantity of interstellar material.

Sweeping the Milky Way with binoculars or a low-power telescope will reveal so many stars that to count them by ordinary methods would take more than a lifetime. The belt is fairly well defined, and its stars seem to be bunched closely together, giving an impression of extreme over-crowding. But the universe is not a crowded place, and the stars in the Milky Way are no more packed than those in the rest of the sky. The luminous band itself is nothing more than a line-of-sight effect, due to the way in which our star-system or Galaxy is shaped.

A rough diagram of the Galaxy is given in Fig. 61. The stars are arranged in a form which bears some resemblance to two plates clapped together by their rims, with the Sun (S) well away from the centre. The dimensions of the

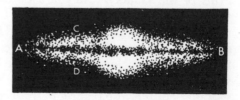

Fig. 61. Position of the Sun (S) in the Galaxy. Looking along the main plane (AB) results in the Milky Way effect.

"plate" are known with fair certainty, and the diameter (AB) proves to be 100,000 light-years, with the greatest breadth of it only one-fifth of this. We can now understand the reason for the Milky Way effect. When we look along SA or SB, we see many stars almost in the same line of sight, but when we look along SC or SD there are far fewer objects to be seen.

Actually, it is not possible to see all the way from S to B. In the main plane of the Galaxy there is a great deal of obscuring material, both dust and gas, and starlight cannot penetrate it any more than a car's headlamps can penetrate a thick fog. The centre or nucleus of the galaxy lies in the direction of the rich star-clouds in Sagittarius (Map VIII), and here we have some glorious telescopic fields, but what lies beyond these fields can never be seen. Fortunately, the new science of radio astronomy has come to our rescue. Radio waves are not blocked by the interstellar matter, any more than a man's voice is blocked by fog, and we are at last learning more about the core of our stellar system.

Radio astronomy has also helped us to find out something about the structure of the Galaxy. It proves to be spiral, not unlike a vast Catherine-wheel, and the whole system is in rotation round its centre. The Sun takes about 225 million years to complete one circuit, so that it has been round only once since the far-off times when the Coal Measures were being laid down in the period of Earth history known to geologists as the Carboniferous. But though the Sun is moving round the centre in an almost circular orbit, other stars have paths of different types. It has now been established that there are two distinct "families" of stars, known generally as Populations I and II.

Population I stars, such as the Sun, are found in the spiral arms of the Galaxy. They are of various spectral types, but the most luminous of them are Blue Giants. On the other hand the senior members of Population II are Red Giants of vast size and relatively low temperature. Population II stars are found in the nucleus of the Galaxy, and also penetrate the vacant spaces between the spiral arms, while the stars of globular clusters also belong to this type. Since some Population II objects are revolving more slowly round the nucleus, and in more elliptical orbits, they seem to have high velocity with respect to ourselves, just as a slow-moving push-bicyclist will seem to have "high velocity" relative to a stream of cars which is moving steadily as a group.

It is also interesting to note that in Population II areas,

there is less of the cosmic obscuring matter which is such a nuisance to us. The areas have in fact been "swept clean", but in Population I regions the Blue Giants are always associated with clouds of dust and bright gaseous nebulæ.

It is pointless to say much about methods of observing the Milky Way, except to repeat that a low power is to be preferred unless some particular object such as a faint nebula is to be examined. There are innumerable rich star-fields, particularly in the Cygnus area and in Sagittarius, and one never tires of sweeping about in these glorious regions, even though the chances of making a useful discovery are very small.

Two of the most striking of the objects in southern skies are the Clouds of Magellan, or Nubeculæ. There are two of them, one much more conspicuous than the other, and it used to be thought that they were detached portions of the Milky Way; but it is now known that they are separate star-systems, and they may probably be regarded as "satellites" of our Galaxy. Each is concentrated towards its centre, while in the case of the Large Cloud there are vague indications of a spiral structure (Map XI).

Both Clouds are prominent naked-eye features, and contain hundreds of thousands of observable stars. Fortunately some of these stars are Cepheids, which have made distance estimations possible. In the Large Cloud, Blue Giant stars are common, and there is also much dust and nebulosity, so that the population appears to be mainly of Class I. Here too lies one of the most luminous stars known, the remarkable variable S Doradûs, which is the equal of a million Suns and yet is so far away that we cannot see it at all except by using a telescope. It would take many pages to describe all the various features to be found in the Clouds; they are superb objects, and it is a great pity that they can never be seen from North America.

Vast though they are, the Clouds are far smaller than the system in which we live. Until recently it was indeed thought that the Milky Way was the largest of all galaxies, so that it had a special status in the universe, but this is far from being the case. There are millions of galaxies within range of our telescopes, and there is no longer any reason to suppose that our own system is of exceptional size.

There is still a tendency to refer to the galaxies as "spiral nebulæ", but this is a bad term. Not all the galaxies are spiral, and certainly none of them is a nebula in the proper sense of the word, though they do contain nebulæ of the same type as the Sword of Orion, as well as open clusters and globulars.

The nearest of the major galaxies, M.31, is easily found, since it can be seen with the naked eye. It lies in Andromeda, and is shown in Map VII. For many years it was thought to lie inside our own system, even though its spectrum showed that it was made up of stars and was not a luminous gas-cloud; but there were also suspicions that it might lie outside the Milky Way altogether. The riddle was solved in 1923, when Cepheids were discovered in the spiral arms. These Cepheids proved to be so remote that M.31 could no longer be regarded as a member of our Galaxy, and a first estimate of its distance gave a value of 900,000 light-years.

Other Cepheids were found, and in 1944 a correction was made which reduced the distance of the galaxy to 750,000 light-years. All seemed to be well, but in 1952 stellar astronomers had a rude shock. It was found that the Cepheid scale was badly in error, because the difference between Population I and Population II Cepheids had not been realized; the result was that the whole distance-scale of the outer universe had to be doubled. Instead of lying at a mere 750,000 light-years, the Andromeda Galaxy was a million and a half light-years away. Further investigations have increased this still more, and the latest estimate is 2,200,000 light-years. The light now entering our eyes started on its journey towards us before the beginning of the last Ice Age.

It also followed that instead of being smaller than the Milky Way, the Andromeda Galaxy is larger, with a total mass at least $1\frac{1}{2}$ times as great. It too is in rotation; it too has its Population I stars (mostly in the arms) and Population II stars (mainly in the nucleus), as well as globulars, open clusters, gaseous nebulæ and even two satellite galaxies of the same status as our Clouds of Magellan. Novæ have been observed, and in 1885 there was even a supernova which flared up to the sixth magnitude. At maximum it could only just be seen without a telescope, but had it appeared inside our own Galaxy it would probably have shone in our skies more brilliantly than Venus.

The Andromeda Galaxy is spiral, but as it is not face-on to us the spectacular Catherine-wheel effect is largely lost. I have always regarded it as rather a disappointing object in a small telescope, and in a 3-inch refractor it looks like a badly-defined patch of mistiness. Powerful telescopes are needed to show it really well, but the best results are obtained by means of photography. In recent years, radio waves emitted by the galaxy have also been detected.

Other galaxies are better presented, so that they appear as Catherine-wheels. In some cases, however, there is no spiral structure. A few galaxies (such as the Small Cloud of Magellan) are virtually formless, while others are elliptical or globular. It used to be thought that the different shapes of galaxies indicated different stages in evolution, but this plausible-sounding idea has now been rejected by most astronomers. There is much that we do not know. For instance, the cause of spiral structure is still very much of a mystery.

Galaxies are apparently most numerous away from the Milky Way zone. This does not mean that there is anything lop-sided about their distribution; the effect is due purely to the obscuring matter near the galactic plane (AB in Fig. 61). There are groups of galaxies here and there, and the total number of known external systems is staggeringly great, though even the giant telescopes of to-day cannot show us more than a small part of the universe.

Spectra of galaxies are not particularly easy to study. Each is made up of the combined spectra of millions of bodies of all types, and the result is bound to be much less clear than with a spectrum of a single star. However, one thing has become clear: nearly all the spectra of galaxies show a red shift, which indicates a velocity of recession.

Apart from the Andromeda Spiral, the fainter spiral in Triangulum, the two Nubeculæ and more than fifteen minor systems which make up our own "local group", all the galaxies are racing away from us, and the more distant they are the faster they go. For instance, the remote galaxy 3C-295 in Boötes seems to be about 5,000 million light-years away, and to be receding at almost half the velocity of light.

If we accept this principle, we must conclude that the whole universe is expanding, with every group of galaxies racing away from every other group. The red shifts do not indicate that our own particular area is any way exceptional; the situation may be visualized by picturing a balloon filled with coloured gas—when the balloon is burst, the gas expands, each part of it receding from each other part. The analogy is admittedly not very accurate, but it is the best that can be done.

There are some astronomers who doubt whether the red shifts in the spectra of galaxies are due to the ordinary Doppler effect. If there is another explanation, the whole idea of an expanding universe might have to be drastically revised. This is not likely, but neither is it impossible—and of late there have been startling developments, due in the main to the new science of radio astronomy.

There are some galaxies which are remarkably powerful radio sources. One such object if Cygnus A, which is so dim optically that giant telescopes are needed to show it at all. It apparently lies at a distance of some 200 million light-years, but even so it is one of the strongest radio sources in the sky, and the cause of this emission is still not known with any certainty at all. A few years ago there was a highly plausible theory, according to which Cygnus A and others of its kind were made up of two galaxies which were in collision, and were "passing through" each other in the manner of two orderly crowds moving in opposite directions; the individual stars would seldom or never suffer direct hits, but the interstellar matter would be colliding all the time, so producing the radio emission. This sounded most satisfactory, but it was then found that the process was hopelessly inadequate to account for the remarkable power of the radio waves, and the whole idea of colliding galaxies had to be cast upon the scientific scrap-heap. Unfortunately, nobody has yet provided a substitute theory which is any better. We do know, however, that there are radio-emitting galaxies which show evidence of having undergone violent explosions in their central parts.

In 1963 there was a new development. By then, many radio sources had been tracked down, and identified either with galaxies, supernova remnants or other known objects. However, some sources seemed to coincide in position with stars. One, in particular—the source known by its catalogue number of 3C-273—coincided with what seemed like a rather faint bluish star, which had been recorded photographically often enough. The whole situation seemed peculiar, and when the radio astronomers asked the American optical astronomers to take a closer look at the spectrum of the "star", some amazing facts emerged. The main surprise was that the bluish object was not a star at all. It had a totally different kind of spectrum, and a tremendous red shift, which presumably meant that it was very remote and was receding very rapidly. This was the first-identified of the objects now known as quasi-stellar objects or, more commonly, as quasars.

The first quasar would have to lie at a distance of about 1,500,000,000 light-years, assuming its red shift to be a normal Doppler effect. Yet it was stellar in appearance—and the final result was that if it were as remote as this, and shining with the magnitude measured by ordinary visual methods, the luminosity must be equal to 100 whole galaxies put together! Since the quasar was clearly much smaller than a galaxy, in view of its starlike aspect, this conclusion seemed to make no sense at all. Nobody could imagine how so small an object could be radiating so much energy; and let it be admitted that the problem is still as baffling as ever. Other quasars were soon identified, some of them still farther away than 3C-273 and receding even more rapidly.

Quasars are, on the whole, the most remarkable objects ever found in the sky. If they are as distant as their red shifts indicate, they must be drawing upon some energy-source about which we know nothing at all. All sorts of ideas have been put forward, such as the production of energy from the gravitational collapse of what may be termed a "super-star", but we are still very much in the dark, and all we can do is to await the results of further research.

These are fascinating problems indeed, but to discuss them at all fully would be beyond the scope of a book devoted to the needs of the amateur astronomer equipped with a modest telescope. Yet the detection of the quasars shows us how little we really know, and it is not impossible that our whole idea of the make-up of the universe may have to be altered during the next few years.

Chapter Eighteen

BEGINNINGS AND ENDINGS

ONLY A FEW centuries ago, the world was believed to be of fairly recent formation. Archbishop Ussher of Armagh summed matters up in 1654, when he stated categorically that the Earth came into being at nine o'clock in the morning of October 26, 4004 B.C. Nowadays we know that the problem is not so simple as this, and that the Earth is well over four thousand million years old, while the Sun is presumably older still.

The Sun must have been formed from material in the Galaxy, but when we come to consider how the universe itself was created we run up against a blank wall. There are plenty of theories, but most of them start from the assumption that matter was created at a definite moment in time, which is not particularly helpful in view of the fact that we do not really understand what "time" is. The Belgian mathematician Lemaître believed that the universe was once concentrated in a single giant radioactive atom, and that time and space began when this blew up; according to other theories, the original universe consisted of a mass of diffuse gas distributed uniformly throughout all space. Whether we adopt the "big bang" or the quieter creation, we are still none the wiser. However far back we go, we can always picture a still earlier period. The only way to try to solve the problem is to use the language of very abstruse mathematics, but even this leads us into a blind alley.

At all events, there must have been a time when there were no galaxies. Presumably the galaxies condensed out of the widely-spread material, and in turn the stars condensed out of the galaxies. We are still very uncertain about the exact way in which a star is born, but there is little doubt that nebular matter is responsible, and it is possible that some of the stars inside the Orion nebula are true celestial infants.

When we come nearer home, we are slightly more confident of our facts. However the planets were formed, the Sun was in some way responsible. It used to be thought that the bodies of the Solar System were thrown-off pieces of the Sun itself, probably drawn off by the tidal pull of a passing star, but this idea has now been abandoned, so that an alternative mode of formation must be sought. The Cambridge astronomer Hoyle once suggested that the planets are the result of the supernova disruption of a former binary companion of the Sun, whereas according to Von Weizsäcker the material of the Solar System was picked up by the Sun during its passage through an interstellar "cloud". At the moment Von Weizsäcker's theory seems to meet with general approval, but there is no guarantee that it is correct.

So much for the past; but what lies ahead? Will the universe last for ever, or will it finally die, so that nothing remains but dead, lifeless bodies scattered through space?

Here again we have to admit that we simply do not know. If we suppose the universe to be eternal, then we have to picture a period of time which has no ending; if not, then we must concede that "time" itself comes to an end, which is equally beyond our mental powers.

In the late 1940's a group of astronomers at Cambridge, headed by H. Bondi and T. Gold, advanced a new and daring idea. They supposed that the universe has always existed, and will exist for ever; as old stars and galaxies die, matter is spontaneously created out of nothingness, so that new galaxies can be produced. Of course, there was no suggestion that a fresh galaxy would suddenly appear in recognizable form. The rate of creation of new matter would be too slow to be detectable, any more than it would be possible for us to detect a new sand-grain in the whole of the Sahara Desert.

According to this theory, modified later by F. Hoyle, the universe would be in a steady state, and must always have looked much the same as it does now; there would be the same numbers of galaxies, even though the individual galaxies would not always be the same. It is an attractive idea, but unfortunately it has not stood up to careful investigation.

Consider the two rival theories—the evolutionary or "big bang", and the steady-state. On the first hypothesis, the matter in the universe was once more closely packed than it is now; on the second, the average distribution has always been the same as at present. If we could go back in time, and see the universe as it used to be thousands of millions of years ago, we would have something definite to guide us. Closer packing of the galaxies would favour evolution; no change in the distribution of the galaxies would support the steady-state idea.

We cannot achieve time-travel, but we can do something almost as good. When we examine a galaxy at a distance of (say) 5,000 million light-years, we are seeing it as it used to be 5,000 million years ago; in other words, we are looking back into the past. The method, then, is to study the distribution of very remote galaxies, and see whether they are more closely crowded than the systems nearer at hand.

Optical telescopes cannot reach out far enough, but radio telescopes can probe farther. Work by Sir Martin Ryle at Cambridge has shown that the distribution of remote galaxies is *not* the same as with the closer parts of the universe, and it now seems that the steady-state theory must be given up, at least in its classical form. It has now been rejected by almost all authorities, albeit with considerable reluctance.

On the other hand, this does not prove that the universe began with a "big bang" and is now evolving toward eventual death. It has been suggested that the universe is in an oscillating condition, and that the cycle has been repeated many times. At present, the galaxies are in a state of spreading-out (assuming, of course, that we accept the evidence of the red shifts in their spectra), but it may be that in the future this mutual recession will cease, and that the galaxies will come together once more before embarking upon a new phase of expansion. But we are still uncertain of our ground, and the discovery of the quasars has shown again how little we really know. Quasars, indeed, may provide us with vital clues, whether they really are immensely remote and super-luminous (as is the official view so far) or whether they will turn out to be relatively local. Far from saying the last word, we cannot be sure that we have said even the first.

It seems that our own galaxy, at least, must die; but men of the Earth will have vanished from the scene long before the end of the story. The Sun will eventually become more luminous, and there must come a time when our world will be too hot to support life. Even if it is not destroyed, it will become a scorched globe devoid of air and water, and all living creatures will have perished from its surface.

There is no immediate danger. The crisis will not come for at least eight thousand million years, and by that time conditions will in any case be so different that it is pointless to speculate about them. Mankind may have destroyed itself by atomic warfare; it may simply have died out, just as the great reptiles vanished over seventy million years ago; or it may be so altered in form that to ourselves it would seem wholly alien. At best, it may have learned so much that our remote descendants will be able to save themselves by abandoning their home world and migrating to another planet.

Of course, this is nothing more than fantasy. Our brains are not able to appreciate such a time-span, and we must accept our limitations. We are creatures of the present; the universe in which we live is spread out for inspection, and everybody can play a part, from the observer who photographs galaxies with the Palomar reflector down to the humble amateur who studies the Moon with the aid of a portable telescope set up in his back garden.

Appendix I

PLANETARY DATA

Planet	Distance from Sun, in millions of miles			Sidereal Period	Synodic Period days	Axial Rotation (Equatorial)
	Max.	Mean	Min.			
MERCURY	43	36	29	88 days	115·9	58d. 5h.?
VENUS	67·6	67·2	66·7	224·7 „	583·9	247d.?
EARTH	94·6	93·0	91·4	365 „	—	23h. 56m.
MARS	154·5	141·5	128·5	687 „	779·9	24h. 37m. 23s.
JUPITER	506·8	483·3	459·8	11·86 years	398·9	9h. 50m. 30s.
SATURN	937·6	886·1	834·6	29·46 „	378·1	10h. 14m.
URANUS	1,867	1,783	1,699	84·01 „	369·7	10h. 48m.
NEPTUNE	2,817	2,793	2,769	164·79 „	367·5	About 14h.?
PLUTO	4,566	3,666	2,766	247·70 „	366·7	6d. 9h.

Planet	Diameter in miles (equatorial)	Apparent Diameter seconds of arc		Maximum Magn.	Axial Incl. Degrees	Mass	Vol.
		Max.	Min.			Earth = 1	
MERCURY	2,900	12·9	4·5	−1·9	?	0·04	0·06
VENUS	7,700	66·0	9·6	−4·4	?	0·83	0·88
EARTH	7,927	—	—	—	23·5	1	1
MARS	4,200	25·7	3·5	−2·8	25·2	0·11	0·15
JUPITER	88,700	50·1	30·4	−2·5	3·1	318	1,312
SATURN	75,100	20·9	15·0	−0·4	26·7	95	763
URANUS	29,300	3·7	3·1	+5·6	98	15	50
NEPTUNE	27,700	2·2	2·0	+7·7	29	17	43
PLUTO	About 8,000	0·3(?)	0·2(?)	< 13	?	?	?

SATELLITE DATA

Satellite	Mean dist. from centre of primary Thousands of miles	Sidereal Period d.	h.	m.	Diameter, miles	Maximum Mag.
EARTH						
Moon	239	27	7	43	2,160	—12·5
MARS						
Phobos	5·8		7	39	10	10
Deimos	14·6	1	6	18	5	11
JUPITER						
Amalthea (V)	113		11	57	150	13
Io (I)	262	1	18	28	2,310	5·5
Europa (II)	417	3	13	14	1,950	5·7
Ganymede (III)	666	7	3	43	3,200	5·1
Callisto (IV)	1,170	16	16	32	3,220	6·3
Hestia (VI)	7,120	250	16		100	13·7
Hera (VII)	7,290	259	16		35	17
Demeter (X)	7,300	260	12		15	18·8
Adrastea (XII)	13,000	*625			14	18·9
Pan (XI)	14,000	*700			19	18·4
Poseidon (VIII)	14,600	*739			35	16
Hades (IX)	14,700	*758			17	18·6
SATURN						
Janus	98		17	58	150(?)	14
Mimas	113		22	37	300	12·1
Enceladus	149	1	8	53	400	11·6
Tethys	183	1	21	18	800	10·6
Dione	235	2	17	41	1,000	10·7
Rhea	328	4	12	25	1,100	9·7
Titan	760	15	22	41	3,500	8·2
Hyperion	920	21	6	38	200	13·0
Iapetus	2,200	79	7	56	2,000(?)	9
Phœbe	8,050	*550	10	50	150	14

* Denotes retrograde motion.

Satellite	Mean dist. from centre of primary Thousands of miles	Sidereal Period d. h. m.			Diameter, miles	Maximum Mag.
URANUS						
Miranda	76	*1	9	50	200	17
Ariel	119	*2	12	29	1,500	14
Umbriel	166	*4	3	28	800	14·7
Titania	272	*8	16	56	1,500	14
Oberon	364	*13	11	7	1,500	14
NEPTUNE						
Triton	220	*5	21	3	3,300	13
Nereid	3,500	359			200	19·5

* Denotes retrograde motion. The diameters and magnitudes of the fainter satellites are most uncertain, and different authorities give different values. For instance, the following diameter estimates are adopted in the *Handbook of the British Astronomical Association*: Io 2,000 miles, Europa 1,800, Ganymede 3,100, Callisto 2,800, Titan 3,000, Triton 2,300, Iapetus only 700. All we can really say is that, generally speaking, the values are of the right order.

Appendix III

MINOR PLANET DATA

THE FOLLOWING list includes data for the first ten minor planets to be discovered. Objects with interesting orbits, such as the Trojans and the "Earth-grazers", are in general too faint to be seen with amateur-owned equipment.

Number	Name	Diameter, miles	Sidereal Period years	Mean Dist. from Sun, millions of miles	Orbital Incl., deg. min.		Max. Mag.
1	CERES	427	4·60	257·0	10	36	7·4
2	PALLAS	280	4·61	257·4	34	48	8·7
3	JUNO	150	4·36	247·8	13	00	8·0
4	VESTA	241	3·63	219·3	7	08	6·0
5	ASTRÆA	111	4·14	239·3	5	20	9·9
6	HEBE	106	3·78	225·2	14	45	8·5
7	IRIS	93	3·68	221·5	5	31	8·7
8	FLORA	77	3·27	204·4	5	54	9·0
9	METIS	135	3·69	221·7	5	36	8·3
10	HYGEIA	220	5·59	292·6	3	49	9·5

Appendix IV

ELONGATIONS AND TRANSITS OF THE INFERIOR PLANETS

(1) MERCURY

Eastern elongations (evening star)

1964   April 7, August 5, November 30
1965   March 21, July 18, November 13
1966   March 5, June 30, October 26
1967   February 16, June 12, October 9
1968   January 31, May 24, September 20
1969   January 13, May 5, September 3, December 27
1970   April 18, August 16, December 10

Western elongations (morning star)

1964   January 26, May 24, September 18
1965   January 8, May 6, September 2, December 21
1966   April 18, August 16, December 4
1967   March 31, July 30, November 17
1968   March 13, July 11, October 31
1969   February 23, June 23, October 14
1970   February 5, June 5, September 28

(2) VENUS

	Greatest elonga-tion, E.	Inferior conjunction	Greatest elonga-tion, W.	Superior conjunction
1964	Apr. 10	June 19	Aug. 29	—
1965	Nov. 15	—	—	Apr. 12
1966	—	Jan. 26	Apr. 6	Nov. 9
1967	June 20	Aug. 29	Nov. 9	—
1968	—	—	—	June 20
1969	Jan. 26	Apr. 8	June 17	—
1970	Sept. 1	Nov. 10	—	Jan. 24

TRANSITS OF MERCURY will occur on 1970 May 9, 1973 November 9, 1986 November 12, and 1999 November 14.

TRANSITS OF VENUS will occur on 2004 June 7 and 2012 June 4, after which there will be no more until 2117 December 10 and 2125 December 8.

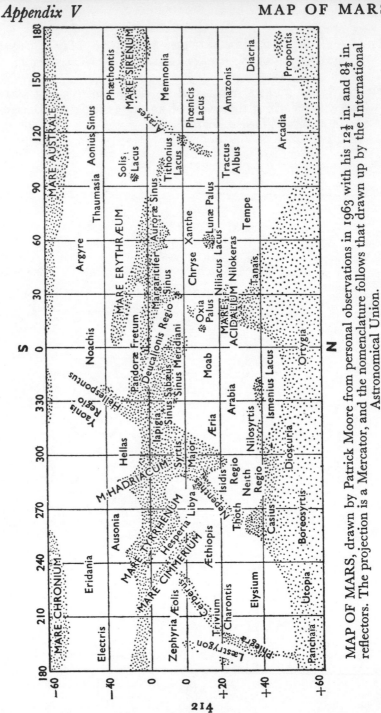

MAP OF MARS, drawn by Patrick Moore from personal observations in 1963 with his 12½ in. and 8¾ in. reflectors. The projection is a Mercator, and the nomenclature follows that drawn up by the International Astronomical Union.

Many maps of Mars have been produced, some of which show a great amount of detail. The official nomenclature is that of the map compiled by the International Astronomical Union*, which shows, among other features, various canals.

The map given here is based upon my own observations, made during 1963. The opposition of 1963 was, of course, rather unfavourable, but at least Mars was well north of the equator; the planet's northern hemisphere was tilted toward us. The polar cap was much in evidence, and there were various short-lived cloud phenomena here and there on the disk.

I do not claim that this map is of extreme precision; it is not intended to be. What I have done is to put in the features that I was able to observe personally, making the positions as accurate as possible, and taking care to omit everything about which I was not fully satisfied. No Lowell-type canals are shown, for the excellent reason that I have never seen any. Broad features such as the Nepenthes-Thoth cannot be classed as canals in the accepted sense of the word.

Some of the Martian markings are very easy to observe. In 1963 I was able to see various dark features with a 3in. refractor, without the slightest difficulty; most prominent of all are the Syrtis Major in the southern hemisphere and the Mare Acidalium in the northern, though Sinus Sabæus, Mare Tyrrhenum and other dark areas are also clear. The more delicate objects require larger apertures; I doubt whether, in 1963, Solis Lacus could have been glimpsed with anything less than an 8½in. reflector, though it is of course possible that a 6in. would have shown it to observers with keener eyes than mine.

Obviously, not all the features shown here are visible at any one moment. The map was compiled from more than fifty separate drawings made at different times. Elusive features which were marked as "suspected only" have been omitted. Since the chart is drawn to a Mercator projection, the polar regions are not shown, but this does not much matter—the south pole was badly placed (indeed, the actual pole was tilted away from the Earth), while the north polar region was covered with its usual white cap, which shrank steadily as the Martian season progressed.

No doubt many observers using the same instruments would have seen more than I did. All I will claim is that my rough map, unlike some other published charts, does not show any features which are of dubious reality!

* Reproduced in *Practical Amateur Astronomy*, uniform with the present book, in which the Mars chapter is by M. B. B. Heath.

Appendix VI

OPPOSITIONS OF PLANETS, 1960–70

(I) MARS

Year	Opposition date	Max. apparent diameter	Max. magnitude
1960	December 30	15·4	−1·3
1963	February 4	14·0	−0·9
1965	March 9	14·0	−0·9
1967	April 15	15·6	−1·3
1969	May 31	19·3	−2·0

(II) JUPITER

1960	June 19	1965	December 18
1961	July 25	1967	January 20
1962	August 31	1968	February 20
1963	October 8	1969	March 21
1964	November 13	1970	April 21

The maximum magnitude at opposition is only slightly variable, and ranges from −2·5 in 1963 to −2·0 in 1969 and 1970.

(III) SATURN

1960	July 7	1965	September 6
1961	July 19	1966	September 19
1962	July 31	1967	October 2
1963	August 13	1968	October 15
1964	August 24	1969	October 28
		1970	November 11

The maximum magnitude at opposition is somewhat variable, owing to the differing presentation of the ring system. It will be +0·1 in 1969, but was only +0·8 in 1965 and 1966.

JUPITER: TRANSIT WORK

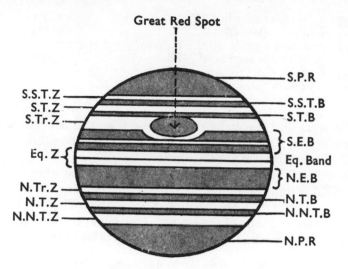

Nomenclature of Jupiter

THE DIAGRAM shows the main belts and zones:

SPR = South Polar Region
SSTZ = South South Temperate Zone
SSTB = South South Temperate Belt
STZ = South Temperate Zone
STB = South Temperate Belt
STrZ = South Tropical Zone
SEB = South Equatorial Belt
Eq. Z = Equatorial Zone
Eq. Band = Equatorial Band
NEB = North Equatorial Belt
NTrZ = North Tropical Zone
NTB = North Temperate Belt
NTZ = North Temperate Zone
NNTB = North North Temperate Belt
NNTZ = North North Temperate Zone
NPR = North Polar Region

The following is a typical extract from my own observation diary:

1963 November 4, 12½-inch reflector. Conditions very variable.

GMT	Feature	Longitude System I	System II	Remarks
19.57	c. of white spot in STZ		182·1	× 360
20.06	f. of this white spot		187·5	
20.21	c. of white patch on the Equator	253·4		
20.22	p. of visible section of NTB		197·2	
20.30	f. of the white patch on the Equator	258·9		
20.37	f. of dark mass on N edge of NEB	263·1		

(c = centre, p = preceding, f = following)

To work out the longitudes of the features, use the tables given in the *B.A.A. Handbook,* which give the longitude of the central meridian for various times.

Example. The 19.57 transit of the centre of the white spot in the STZ. From the tables: longitude of the central meridian (System II) for 16h on November 4 is 038·9. This is 3h 57m earlier than the time of the transit. Therefore, the longitude for 19.57 may be worked out from the second table in the *Handbook*:

Long at 16h. Nov. 4:	=	038·9
+ 3h	=	108·8
+ 50m	=	30·2
+ 7m	=	4·2
= + 3h 57m	=	182·1

If the calculated longitude works out at over 360, then subtract 360°.

It is important to use the correct System; System I is bounded by the N edge of the SEB and the S edge of the NEB, all the rest of the planet being System II. If the wrong tables are used, the results can be very peculiar indeed, since Systems I and II differ by many degrees.

Even more convenient tables are given in B. M. Peek's admirable book, *The Planet Jupiter,* and the example may then be worked as follows:

Long. at 16h, November 4 (from the *Handbook*)	=	038·9
+ 3h 57m (from Peek's tables)	=	143·2
		182·1

SATURN: INTENSITY ESTIMATES

DEFINITE FEATURES on the disk of Saturn are so rare that our knowledge of the rotation periods of the different zones is not nearly so complete as in the case of Jupiter. Valuable work can however be done in estimating the brightness of the different zones, as well as of the rings, as these are suspected of variation. The scale adopted is from a value of o (brilliant white) to 10 (black shadow). In general, Ring B is the brightest feature, and the outer part has a brightness of 1.

The easiest way of recording is to prepare a sketch (perhaps a rough one) of the globe and rings, and then merely jot down the numerical values upon the drawing itself. It is best to make each estimate twice; first, start from the darkest feature and work through to the lightest, then begin once more, this time with the lightest feature, which is almost always the outer part of Ring B. The following is an extract from my own notebook:

1956 May 21, oh. to oh. 20m. 12½-in. Refl. × 460. Conditions good.

Ring B, outer	= 1½	N.E.B. intermediate zone	= 5½
Ring B, inner	= 2½	N. Equatorial Belt, N.	
Equatorial Zone	= 3	component	= 6½
N. Temperate Zone	= 4½	Encke's Division	= 7
N. Polar Region	= 5	Ring C	= 7
N. Temperate Belt	= 5½	Cassini's Division	= 7½
Ring A	= 6½	Shadow, Rings on Globe	= 8
N.E. Equatorial Belt, S.		Shadow, Globe on Rings	= 8½
component	= 6½		

Of course, the different parts of the ring-system cannot be seen individually when the rings are edge-on to us, as in 1966.

Appendix IX

RECENT AND
FORTHCOMING ECLIPSES

(1) LUNAR ECLIPSES

IN THE TABLE, an asterisk denotes that the eclipse is total. The last two columns show whether or not the eclipse can be seen in England and in the U.S.A.: "partly" may mean that the Moon is very low in the sky, or that the Moon rises or sets while the eclipse is in progress.

Year	Date	Time of mid-eclipse, h.	Percentage of Moon eclipsed	Visible in England	U.S.A.
1964	June 25	1·07	*	Yes	Partly
1964	December 19	2·35	*	Yes	Yes
1965	June 14	1·51	20	Yes	Partly
1967	April 24	12·07	*	No	Partly
1967	October 18	10·16	*	No	Partly
1968	April 13	4·49	*	Partly	Yes
1968	October 6	11·41	*	No	Partly
1970	February 21	8·31	5	No	Yes
1970	August 17	3·25	40	Partly	Yes
1971	February 10	7·42	*	Partly	Yes
1971	August 6	19·44	*	Partly	Yes
1972	January 30	10·53	*	No	Yes
1972	July 26	7·18	60	No	Yes
1973	December 10	1·48	10	Yes	Yes
1974	June 4	22·14	80	Yes	No
1974	November 29	15·16	*	No	No
1975	May 25	5·46	*	No	Yes
1975	November 29	22·24	*	Yes	Partly

(II) SOLAR ECLIPSES 1964–1972

Year	Date	Time of mid-eclipse, h	Remarks
1964	June 10	4·23	Partial
1964	July 9	11·30	Partial
1964	December 4	1·19	Partial
1965	May 30	21·14	Total: Pacific only
1965	November 23	4·11	Annular in Borneo, New Guinea, etc.
1966	May 20	9·43	Partial in Britain; annular over parts of Greece.
1966	November 12	14·27	Total in the N. Pacific, parts of South America
1967	May 9	14·57	Partial
1967	November 2	5·48	Total in Antarctica
1968	March 28	22·48	Partial
1968	September 22	11·9	Total in parts of North Russia
1969	March 18	4·52	Annular in the Indian Ocean, New Guinea, etc.
1969	September 11	19·56	Annular in North Pacific, parts of South America
1970	March 7	17·43	Total in Central Pacific, Mexico, etc.
1970	August 31	22·2	Annular in South Pacific
1971	February 25	9·49	Partial
1971	July 22	9·15	Partial
1971	August 20	22·54	Partial
1972	January 16	10·53	Annular in Antarctica
1972	July 10	19·39	Total in Japan, Canada, parts of the Atlantic

The next total eclipses visible as such in Britain will take place in 1999 August 11, 2090 September 23, and 2135 October 7.

Appendix X

ARTIFICIAL SATELLITES

THE FIRST artificial satellite, Sputnik I, was launched by the Russians on October 4,1957. Many others have followed it; one need only mention the Tiros weather satellites, the plastic balloons Echo I and II, and of course Telstar, which ushered in a new era so far as communications were concerned.

The real importance of a satellite cannot be judged from its brightness. Echo, a "reflector" of virtually no mass and devoid of instrumentation, looks like a brilliant star crawling across the sky; I have lost count of the number of letters I have had asking me about it. On the other hand Telstar, which opened the way for the first direct television link between Europe and America, was a dim telescopic object. The Tiros weather satellites, which have carried out invaluable scientific work, are also very inconspicuous visually.

There are three basic methods of tracking a satellite: radio, radar and visual. Amateur work by radio and radar methods was very useful in the first stages of the satellite programme, and can still be of some value, though it has now been largely superseded by the efficient global network of American stations. However, it requires complex equipment, and most would-be satellite observers will probably prefer to concentrate upon visual methods.

There is much work to be done here. The orbits of the satellites are always changing, because all the vehicles are affected by air-drag (Echo I, for instance, is a long way out, but its orbit is markedly affected by light-pressure from the Sun). The only way to keep track of the satellites is to check their positions constantly, so that the orbital elements may be adjusted and predictions kept up to date.

What has to be done, then, is to make an accurate estimate of the satellite's position, making sure that the timing is correct to one-tenth of a second or so. A split-action stop-watch is a virtual necessity. Stars are used as reference points; the moving satellite may be timed at the moment when it passes exactly half-way between two identifiable stars, or perhaps when it makes an equilateral triangle with two stars. It is not often that a satellite occults a star, though of course it does happen sometimes.

Bright satellites may naturally be tracked with the naked eye and some, such as Echo II—launched in January 1964— are very conspicuous; but most require optical aid. A wide field is needed; a 2-inch "elbow telescope" is suitable for the more conspicuous telescopic objects, while most amateur tracking stations are equipped with 4-inch reflectors.

It is pointless to go out at night, sweep around the sky and hope to find a moving satellite! For telescopic satellites, predictions must be obtained; these are supplied to serious observers by the B.A.A. Artificial Satellite Section. Photography may be used, though vehicles fainter than the 3rd magnitude are hard to record satisfactorily.

Visual tracking of satellites is an important branch of research, and amateur observers are able to make major contributions to it. Probably the best equipment for the satellite observer is a good pair of binoculars.

Appendix XI

THE LIMITING LUNAR DETAIL
VISIBLE WITH DIFFERENT APERTURES

THE FOLLOWING information is based on work by E. A. Whitaker, formerly Director of the Lunar Section of the British Astronomical Association, and given in the Sectional journal, *The Moon* (Vol. 4, No. 2, page 42; December 1955). The table gives the approximate diameters of the smallest craters half-filled with shadow, and of the narrowest black line certainly distinguishable. Perfect seeing conditions and first-class optical equipment are assumed.

Aperture of O.G. in ins.	*Smallest crater*	*Narrowest Cleft*
1	9 miles	0·5 mile
2	4·5 ,,	0·25 ,,
3	3 ,,	0·16 ,,
4	2·25 ,,	220 yards
6	1·5 ,,	150 ,,
8	1·1 ,,	110 ,,
10	0·9 ,,	90 ,,
12	0·75 ,,	70 ,,
15	0·6 ,,	60 ,,
18	0·5 ,,	50 ,,
33	500 yards	30 ,,

The smallest craterlet that I personally have recorded is probably that on the summit of a mountain peak near the crater Beer. The instrument used was the Meudon 33-inch, and the diameter of the summit depression cannot have been much more than 500 yards. Of course, very much smaller features are shown on the photographs taken from rocket probes.

THE LUNAR MAPS

THESE OUTLINE maps have been constructed from two photographs. The whole lunar surface is covered, but the method has two disadvantages. First, the formations near the eastern and western limbs are under high light, and are consequently not well seen. Petavius, for instance, in the southwest, is really a majestic crater 100 miles across, and when anywhere near the terminator it is a magnificent object, but under this lighting it is hard to make out at all. Secondly, the photographs were taken when the Moon was at favourable libration for the west, so that the western limb regions are shown slighly better than the eastern. ("East" and "west" are used in the classical or astronomical sense.

These defects would be serious for a detailed map, but are not important for the present purpose. The observer may compare the map with the photograph given on the opposite page, and it will be easy to recognize the various formations. Once this has been done, serious work can be commenced; after a while, the observer will be able to identify the craters at a glance.

Only a few features are named on these charts; the remaining names will be found on more detailed maps.

The notes given here are, of course, extremely brief; they refer only to objects which are useful for "landmark" purposes, and to one or two features of particular interest, such as Linné, the Alpine Valley and the Straight Wall.

Names of the Lunar "Seas"

Latin	*English*
OCEANUS PROCELLARUM	OCEAN OF STORMS
MARE IMBRIUM	SEA OF SHOWERS
MARE FRIGORIS	SEA OF COLD
MARE HUMORUM	SEA OF HUMOURS

MARE VAPORUM	SEA OF VAPOURS
MARE FŒCUNDITATIS	SEA OF FERTILITY
MARE TRANQUILLITATIS	SEA OF TRANQUILLITY
MARE SERENITATIS	SEA OF SERENITY
MARE CRISIUM	SEA OF CRISES
MARE HUMBOLDTIANUM	HUMBOLDT'S SEA
MARE SMYTHII	SMYTH'S SEA
MARE AUSTRALE	SOUTHERN SEA
MARE UNDARUM	SEA OF WAVES
MARE SPUMANS	FOAMING SEA
LACUS SOMNIORUM	LAKE OF DREAMS
PALUS SOMNII	MARSH OF DREAM
SINUS IRIDUM	BAY OF RAINBOWS
SINUS RORIS	BAY OF DEW
SINUS ÆSTUUM	BAY OF HEATS
SINUS MEDII	CENTRAL BAY
MARE MARGINIS	MARGINAL SEA

THE WESTERN HALF OF THE MOON (Plate XV; Map 1)

(1) North-Western Quadrant

In this quadrant lie three major seas, the Maria Serenitatis, Tranquillitatis and Crisium, with parts of the Mare Vaporum and the Mare Frigoris, as well as the northern part of the Mare Fœcunditatis. Mare Serenitatis is the most conspicuous, and is one of the best-defined of the lunar seas. On its surface are only two objects of importance, the 12-mile bright crater Bessel and the famous (or infamous) Linné, which used to be described as a deep crater, but is now seen in small telescopes as a mere white spot. A number of ridges cross the Mare Serenitatis. Mare Tranquillitatis is lighter in hue; between it and Serenitatis there is a strait upon which lies the magnificent 30-mile crater Plinius, which has two interior craterlets near its centre.

Of the mountain ranges, the most important are those bordering the Mare Serenitatis: the Hæmus and the Caucasus Mountains, with peaks rising to 8,000 and 12,000 feet respectively. Part of the Alps can also be seen, cut through by the strange Alpine Valley. This is an interesting formation, by far the most conspicuous of its type. South of the Hæmus, and north of the large crater Hipparchus, can be seen the two important clefts of the Mare Vaporum: that of Hyginus (which is basically a crater-chain) and Ariadæus. Each can be seen with any small telescope when near the terminator.

Close to the western limb can be seen two very foreshortened seas, the Maria Smythii and Marginis, while the Mare Humboldtianum lies further north. These can only be well seen under favourable libration.

ARAGO. Diameter 18 miles. It lies on the Mare Tranquillitatis, and has a low central elevation. Near it are several of the interesting swellings or "domes".

ARCHYTAS. A bright 21-mile crater on the north coast of the Mare Frigoris. It has a central peak.

ARISTILLUS. Diameter 35 miles, with walls rising to 11,000 feet above the floor. The walls are bright, and there is a central peak. Inside Aristillus are dark patches and streaks, formerly attributed to vegetation, though this theory is now discounted. It and Autolycus form a pair. Though it lies on the N.W. Quadrant, Aristillus is better shown in the next photograph.

ARISTOTELES. A prominent crater 52 miles in diameter. It and Eudoxus form a notable pair.

ATLAS. Forms a pair with Hercules; it lies N. of the Mare Serenitatis. Diameter 55 miles. The walls are much terraced, rising to 11,000 feet. There is much detail on the floor.

AUTOLYCUS. The companion to Aristillus. Autolycus is 24 miles in diameter, and 9,000 feet deep. It too is better shown on the next photograph.

BOSCOVITCH. On the Mare Vaporum. It is a low-walled, irregular formation, recognizable (like its companion Julius Cæsar) by its very dark floor.

BÜRG. A 28-mile crater between Atlas and Aristoteles, with a large central peak on which is a summit craterlet. East of it lies an old plain traversed by numerous clefts.

CASSINI. On the fringe of the Alps. A curious broken formation, shallow, and 36 miles in diameter. It contains a prominent craterlet, A.

CLEOMEDES. A 78-mile crater near the Mare Crisium. It is broken in the E. by a smaller but very deep crater, Tralles.

DIONYSIUS. A brilliant small crater near Sabine and Ritter.

ENDYMION. This 78-mile crater can always be recognized by the darkness of its floor. Patches on the interior seem to vary in hue, and should be watched.

EUDOXUS. The companion to Aristoteles. It is 40 miles in diameter, and 11,000 feet deep.

FIRMINICUS. Closely S. of the Mare Crisium. It has a diameter of 35 miles, and can easily be identified by its dark floor.

GAUSS. A magnificent 100-mile crater, not well shown in the photograph, but very conspicuous when on the terminator.

GEMINUS. Diameter 55 miles. It lies near Cleomedes, and has lofty walls, which are deeply terraced.

GODIN. Diameter 27 miles. It lies near Ariadæus and Hyginus. Closely north of it lies Agrippa, which is slightly larger but somewhat less deep.

HERCULES. The companion of Atlas. It is 45 miles in diameter; the walls are much terraced, and appear brilliant at times. Inside Hercules lies a large craterlet, A.

JULIUS CÆSAR. A low-walled formation in the Mare Vaporum area. Owing to its dark floor, it is easy to recognize at any time.

MACROBIUS. A 42-mile crater near the Mare Crisium, with walls rising to 13,000 feet. There is a low compound central mountain mass.

MANILIUS. A 25-mile crater on the Mare Vaporum, notable because of its brilliant walls.

MENELAUS. Another brilliant crater; 20 miles in diameter, lying in the Hæmus Mountains. Like Manilius, its brightness makes it easy to identify.

POSIDONIUS. A 62-mile plain on the border of the Mare Serenitatis. Adjoining it to the west is a smaller, squarish formation,
Chacornac, and south is Le Monnier, one of the "bays" with a
broken down seaward wall. Using Dr. Steavenson's 30-inch
reflector at Cambridge, I was unable to see any detail inside
Le Monnier; it must be one of the smoothest areas on the
Moon.

PROCLUS. Closely E. of the Mare Crisium. It is one of the most
brilliant formations on the Moon, and is the centre of a ray
system. Diameter 18 miles.

SABINE and RITTER. Two 18-mile craters on the E. border of the
Mare Tranquillitatis. N.E. of Ritter are two small equal
craterlets. This area was photographed in detail by the U.S.
probe Ranger VIII in 1965.

SCORESBY. A very distinct formation 36 miles across, near the
North Pole. It is much the most conspicuous formation in its
area, and is thus very useful as a landmark.

TARUNTIUS. A 38-mile crater S. of the Mare Crisium, with
narrow walls and a low central hill. It is a "concentric
crater", since it contains a complete inner ring.

(2) *South-West Quadrant*

This quadrant is occupied largely by rugged uplands, and
large and small craters abound. The only major seas are the
small, well-marked Mare Nectaris and most of the larger
Mare Fœcunditatis; the Mare Australe, near the limb, is much
less well-defined. The only mountain ranges of note are the
lofty Leibnitz, right on the limb, which contain the highest
peaks on the Moon (30,000 feet according to the latest
measures, most of which have still to be published), and the
much lower Altai.

ALBATEGNIUS. A magnificent walled plain near the centre of the
disk; the companion of Hipparchus. Diameter 80 miles. The
S.W. wall is disturbed by a deep 20-mile crater, Klein.

CAPELLA. A 30-mile crater near Theophilus. It has a very
massive central mountain, topped by a summit craterlet; the
floor of Capella is crossed by a deep valley. It has a shallower
companion, Isidorus.

CUVIER. This forms an interesting group with Licetus and the irregular Heraclitus. Cuvier is 50 miles across, and lies on the terminator in the photograph, not far from the top of the page.

FABRICIUS. A 55-mile crater, not well shown in the photograph owing to the high light. It has a companion of similar size, Metius. Fabricius interrupts the vast ruined plain Janssen.

GUTENBERG. This and its companion Goclenius lie on the highland between the Maria Nectaris and Fœcunditatis. To the north lie some delicate clefts.

HIPPARCHUS. Well shown on the photograph, but it is low-walled and broken, so that it becomes obscure when away from the terminator. It is 84 miles in diameter, and is the companion of Albategnius. Ptolemæus lies closely east of it.

LANGRENUS. An 85-mile crater, with high walls and central mountain. It is a member of the great Western Chain, which extends from Furnerius in the south and includes Petavius, Vendelinus, Mare Crisium, Cleomedes, Geminus and Endymion.

MAUROLYCUS. A deep 75-mile crater, well shown in the photograph; in the far south of the Moon. East of it lies the larger plain Stöfler, which has a darkish floor.

MESSIER. This curious little crater lies on the Mare Fœcunditatis. It and its companion, Pickering, show curious apparent changes in form and size. Extending E. of them is a strange bright ray, rather like a comet's tail.

PETAVIUS. Not well shown on the photograph; but it is a magnificent object when better placed. Closely W. of it is Palitzsch, which is generally described as a valley-like groove; but using very large telescopes, I have found it to be a crater-chain.

PICCOLOMINI. A 56-mile crater S. of Fracastorius and the Mare Nectaris. It is deep and conspicuous, and lies at the W. end of the Altai range.

RHEITA. A 42-mile crater. Associated with it is the famous Rheita Valley. This has been described as a "groove" and attributed to a falling meteor; but it is in fact a crater-chain, and so no such explanation can be admitted. There is another similar formation not far off, associated with the crater Reichenbach. Rheita lies not far from the Mare Australe.

STEINHEIL. This 42-mile crater forms a pair of "Siamese twins" with its similar but shallower neighbour Watt. Near the Mare Australe.

THEOPHILUS. The northern member of the grand chain of which Cyrillus and Catharina are the other members. Theophilus is 65 miles across and 18,000 feet deep; it is one of the most magnificent of the lunar craters. There is a lofty, complex central elevation.

VENDELINUS. One of the Western Chain. It is 100 miles across, but is comparatively low-walled, and is conspicuous only when near the terminator.

VLACQ. One of a group of large ring-plains near Janssen, not far from the Mare Australe. It is 56 miles in diameter, and 10,000 feet deep.

WERNER. A 45-mile crater, shadow-filled in the photograph. It forms a pair with its neighbour Aliacensis, and close by are three more pairs of formations: Apian-Playfair, Azophi-Abenezra, and Abulfeda-Almanon.

WILHELM HUMBOLDT. Not recognizable on the photograph, as it lies on the western limb near Petavius, but it is 120 miles across, with high walls and central mountain, and is magnificent just after Full Moon.

THE EASTERN HALF OF THE MOON (Plate XVI; Map 11)

(1) *North-East Quadrant*

This Quadrant consists largely of "sea"; there is the magnificent Mare Imbrium, with a diameter of 700 miles, as well as parts of the even vaster but less well-defined Oceanus Procellarum, and most of the Mare Frigoris and the Sinus Roris. The chief mountains are the Apennines (certainly the most spectacular on the Moon) and the Jura Mountains, which form part of the Imbrian border. On the limb are the Hercynian Mountains. There are also the lower Carpathians, near Copernicus. Near Full, the most conspicuous objects are Copernicus and Kepler, which are the centres of bright ray systems, and Plato, whose floor is so dark that it can never be mistaken, while Aristarchus is the most brilliant formation on the entire Moon.

ANAXAGORAS. A 32-mile crater not far from the North Pole. It is the centre of a ray system, and is always distinct.

ARCHIMEDES. A 50-mile plain on the Mare Imbrium, with a darkish floor and rather low walls. It forms a superb group with Aristillus and Autolycus.

ARISTARCHUS. The brightest formation on the Moon. Associated with its companion, Herodotus, is the great winding valley, discovered by Schröter. Recent reports of reddish patches in this area are described in the text.

COPERNICUS. The great ray-crater, described in the text.

SINUS IRIDUM. A glorious bay on the border of the Mare Imbrium. When the Sun is rising over it, the rays catch the bordering Jura Mountains, and the bay seems to stand out into the darkness like a handle of glittering jewels.

KEPLER. A 22-mile crater on the Oceanus Procellarum, centre of a very conspicuous system of bright rays. South of it is a crater of similar size, Encke, which is however shallower and is not associated with any bright rays.

OLBERS. A crater on the east limb. It lies N. of Grimaldi, and is 40 miles in diameter. It is not identifiable on the photograph, because of the high light and unfavourable libration; but it is prominent when well placed, and is the centre of a ray system.

PHILOLAUS. A crater near the limb, 46 miles in diameter. It forms a pair with its neighbour Anaximenes. Reddish hues have been reported inside Philolaus, perhaps indicating some unusual surface deposit.

PICO. A splendid 8,000-foot mountain on the Mare Imbrium, S. of Plato, with at least three peaks. Some way S.E. of it is Piton, which is also shown on the first photograph, and has a summit craterlet.

PLATO. This regular, 60-mile formation has a dark floor, and is one of the most interesting features on the Moon. Inside it are some delicate craterlets which show baffling changes in visibility. Plato is always identifiable, and will well repay close and continuous attention.

PYTHAGORAS. A very deep crater 85 miles in diameter, not well shown in the photograph, but magnificent when well placed. There are numerous large formations in this area, and the whole region is in need of attention, as the charts are not as yet completely satisfactory.

STRAIGHT RANGE. A peculiar range of peaks on the Mare Imbrium, near Plato. It is 40 miles long, and the highest mountains attain 6,000 feet.

TIMOCHARIS. A 23-mile crater on the Mare Imbrium, containing a central craterlet. It is the centre of a rather inconspicuous system of rays.

(2) South-East Quadrant

This Quadrant is crammed with interesting features. In the northern part of it lie the well-marked Mare Humorum, part of the Oceanus Procellarum, and most of the vast Mare Nubium; the southern part is mainly rough upland. The chief mountain ranges are the curious low Riphæans, on the Mare Nubium; the Percy Mountains, forming part of the border of the Mare Humorum; and the Dörfels, Rook Mountains, Cordilleras and D'Alemberts, on the limb. The Dörfels attain nearly 30,000 feet, and are thus the highest mountains on the Moon apart from the Leibnitz. It is a great pity that these magnificent ranges are so badly placed.

ALPHONSUS. The great crater close to Ptolemæus. Dark patches may be seen on its floor. It was in Alphonsus that Kozirev, in 1958, reported a visible outbreak of activity. The U.S. vehicle Ranger IX landed in Alphonsus in 1965.

BAILLY. Very obscure on the photograph, but this vast formation is almost 180 miles across, and is thus the largest of the objects generally classed as "craters". It has been aptly described as "a field of ruins". In my 12½-in. reflector, I have seen so much detail inside it that to draw the formation accurately would be excessively difficult.

BILLY. A 30-mile crater S. of Grimaldi. It can be identified at any time because of its very dark floor; it is always distinct. It has a near neighbour, Hansteen, with a much lighter floor.

BIRT. A crater 11 miles in diameter, in the Mare Nubium, near the Straight Wall. It has walls that rise unusually high above the outer plain, and inside it are two of the strange radial bands.

BULLIALDUS. A splendid 39-mile crater on the Mare Nubium, with terraced walls and a central peak. This is one of the most perfect of the ring-plains.

CLAVIUS. Surpassed in size only by Bailly; it is 145 miles across, with walls containing peaks 17,000 feet above the floor. Inside it can be seen a chain of craters, decreasing in size from west to east. When right on the terminator, Clavius can be identified with the naked eye.

CRÜGER. A low-walled crater near Grimaldi, 30 miles in diameter. It can be identified on the photograph by the darkness of its floor, which is rather similar to Billy's.

DOPPELMAYER. An interesting 40-mile bay on the Mare Humorum. The seaward wall can just be traced, and there is a much reduced central mountain.

EUCLIDES. Only 7 miles in diameter, but surrounded by a prominent bright nimbus, well shown on the photographs. It lies near the Riphæan Mountains.

FRA MAURO. One of a group of damaged ring-plains on the Mare Nubium. The other members of the group are Parry, Bonpland and Guericke.

GASSENDI. A magnificent walled plain on the N. border of the

Mare Humorum. It is 55 miles in diameter, and the floor contains a central mountain and numerous delicate clefts. Reddish patches have been seen in and near Gassendi, and are described in the text.

HIPPALUS. Another bay on the Mare Humorum, not unlike Doppelmayer. Near it are numerous prominent clefts, well seen in a small telescope, and there are also clefts on the floor. Near Hippalus is a small crater Agatharchides A, in in which I discovered two radial bands. These bands are useful test objects. I have seen them clearly with an aperture of 6 inches, but keener-eyed observers should detect them with smaller instruments.

GRIMALDI. Identifiable at all times because of its floor, which is the darkest spot on the Moon. It lies close to the east limb. Patches on the floor show interesting variations in hue, and should be watched. Grimaldi has low walls, and is 120 miles in diameter. Near by is a smaller formation, Riccioli, 80 miles in diameter; it too has a very dark patch inside it.

LETRONNE. A bay 70 miles in diameter, lying on the shore of the Oceanus Procellarum not far from Gassendi. There is the wreck of a central elevation.

MAGINUS. A vast walled plain near Clavius and Tycho. It is very prominent when near the terminator, as in the photograph; but it becomes very obscure near Full Moon.

MERCATOR. This and Campanus form a conspicuous pair of craters W. of the Mare Humorum. Each is about 28 miles in diameter, and the only obvious difference between them is that Mercator has a darker floor.

MERSENIUS. A convex-floored 45-mile crater near Gassendi, associated with an interesting system of clefts.

MORETUS. Not well shown on the photograph, but it is a splendid crater 75 miles in diameter and 15,000 feet deep. The central mountain is the highest of its type on the Moon.

PITATUS. Described by Wilkins as being like "a lagoon". It lies on the S. border of the Mare Nubium, and has a dark floor and a low mountain near its centre. It is 50 miles in diameter. East of it is a smaller formation, Hesiodus, and from Hesiodus a prominent cleft runs towards Mercator and Campanus.

PTOLEMÆUS. Over 90 miles across; one of the most interesting formations on the Moon. It lies near the centre of the disk. Its floor is moderately dark. It is the northern member of a chain of three great craters, the other two being Alphonsus and Arzachel. South of this chain lies another, made up of the three formations Purbach, Regiomontanus and Walter.

SCHICKARD. A formation 134 miles in diameter. It can be identified on the photograph, near the S.E. limb, because parts of its floor are darkish. Obscurations have been reported inside it, and it is well worth watching.

SIRSALIS. This and its "Siamese twin", A, lie near the dark-floored Crüger, not far from Grimaldi. Unfortunately they are not identifiable on the photograph. Sirsalis is associated with one of the most prominent clefts on the Moon.

STRAIGHT WALL. The celebrated fault in the Mare Nubium, near Birt. It is shown in the photograph as a white line, but casts considerable shadow before Full, when the illumination is from the reverse direction, so that it then appears as a dark line. Near it are numerous craterlets, some of them visible with very modest apertures. The Wall lies inside a large and obscure ring.

THEBIT. A 37-mile crater near the Straight Wall. It is interrupted by a smaller crater, which is in turn interrupted by a third. The group makes a useful test object for small apertures.

TYCHO. The great ray-crater, described in the text.

VITELLO. A 30-mile crater on the border of the Mare Humorum, with an inner but not quite concentric ring.

WARGENTIN. Most unfortunately, this is not identifiable on the photograph. It lies near Schickard, and is a 55-mile plateau, much the largest formation of its type on the Moon. Though Wilkins and I have used very large apertures to chart over fifty objects on the floor, little detail can be seen in small telescopes; it is nevertheless worth observing. Near Wargentin is an interesting group of craters of which Phocylides is the largest member.

Appendix XIII

SOME OF THE MORE IMPORTANT PERIODIC COMETS

Name	Sidereal Period Years	Distance from Sun, astronomical units	
		Perihelion	Aphelion
ENCKE	3·3	0·3	4·1
GRIGG-SKJELLERUP	4·9	0·9	4·9
TEMPEL II	5·3	1·1	4·9
TUTTLE-GIACOBINI-KRESAK	5·5	1·1	5·1
PONS-WINNECKE	6·2	1·2	5·5
KOPFF	6·3	1·5	5·2
GIACOBINI-ZINNER	6·4	1·0	6·0
SCHWASSMANN-WACHMANN II	6·5	2·1	4·8
D'ARREST	6·7	1·4	5·7
DANIEL	6·7	1·5	5·6
BROOKS II	6·7	1·9	5·4
BORRELLY	7·0	1·4	5·8
FAYE	7·4	1·7	6·0
WHIPPLE	7·5	2·5	5·2
OTERMA	7·9	3·4	4·5
SCHAUMASSE	8·2	1·2	6·9
WOLF I	8·4	2·5	5·8
COMAS SOLÁ	8·6	1·8	6·6
VÄISÄLÄ	10·5	1·8	7·9
TUTTLE	13·6	1·0	10·4
SCHWASSMANN-WACHMANN I	16·1	5·5	7·2
NEUJMIN I	18·0	1·5	12·2
CROMMELIN	27·9	0·7	17·6
STEPHAN-OTERMA	39·0	1·6	20·9
WESTPHAL	61·7	1·3	30·0
BRORSEN-METCALF	69·1	0·5	33·2
OLBERS	69·6	1·2	33·6
PONS-BROOKS	70·9	0·8	33·7
HALLEY	76·0	0·6	35·3

Appendix XIV

SOME OF THE MORE IMPORTANT
ANNUAL METEOR SHOWERS

THIS LIST includes only a few of the many annual showers. The dates given for the beginnings and ends of the showers are only approximate.

Name	Beginning	End	Naked-eye Star near radiant	Remarks
QUADRANTIDS	Jan. 3	Jan. 5	Beta Boötis	Usually a sharp maximum, Jan. 4.
LYRIDS	Apr. 19	Apr. 22	Nu Herculis	Moderate shower. Swift meteors.
ETA AQUARIDS	May 1	May 8	Eta Aquarii	Long paths; very swift.
DELTA AQUARIDS	July 15	Aug. 10	Delta Aquarii	Moderate shower.
PERSEIDS	July 27	Aug. 17	Eta Persei	A rich shower. Meteors very swift.
ORIONIDS	Oct. 15	Oct. 25	Nu Orionis	Moderate shower. Swift meteors.
LEONIDS	Nov. 14	Nov. 17	Zeta Leonis	Not usually a rich shower. Very swift meteors.
ANDROMEDIDS	Nov. 26	Dec. 4	Gamma Andromedæ	Very slow meteors. Very weak shower.
GEMINIDS	Dec. 9	Dec. 13	Castor	Very rich shower.
URSIDS	Dec. 20	Dec. 22	Kocab	Rather weak.

Appendix XV

THE CONSTELLATIONS

IN THE FOLLOWING list, an asterisk indicates that the constellation was listed by Ptolemy; X, that much or all of the constellation is invisible in New York. Zodiacal constellations are distinguished by the letter Z.

Constellations	English Name	Remarks	1st mag. Star or Stars
Andromeda	Andromeda	*	
Antlia	The Air-Pump	X	
Apis	The Bee	X	
Aquarius	The Water-Bearer	*Z	
Aquila	The Eagle	*	Altair
Ara	The Altar	X	
Argo Navis	The Ship Argo	X	Canopus
Aries	The Ram	*Z	
Auriga	The Charioteer	*	Capella
Boötes	The Herdsman	*	Arcturus
Cælum	The Sculptor's Tools	X	
Camelopardus	The Camelopard	–	
Cancer	The Crab	*Z	
Canes Venatici	The Hunting Dogs	–	
Canis Major	The Great Dog	*	Sirius
Canis Minor	The Little Dog	*	Procyon
Capricornus	The Sea-Goat	*Z	
Cassiopeia	Cassiopeia	*	
Centaurus	The Centaur	X	Alpha Centauri, Agena
Cepheus	Cepheus	*	
Cetus	The Whale	*	
Chamæleon	The Chameleon	X	
Circinus	The Compasses	X	
Columba	The Dove	–	

Constellations	English Name	Remarks	1st mag. Star or Stars
Coma Berenices	Berenice's Hair	–	
Corona Australis	The Southern Crown	X	
Corona Borealis	The Northern Crown	*	
Corvus	The Crow	–	
Crater	The Cup	–	
Crux Australis	The Southern Cross	X	Acrux, Beta Crucis
Cygnus	The Swan	*	Deneb
Delphinus	The Dolphin	*	
Dorado	The Swordfish	X	
Draco	The Dragon	*	
Equuleus	The Little Horse	*	
Eridanus	The River Eridanus	*X	Achernar (X)
Fornax	The Furnace	–	
Gemini	The Twins	*Z	Pollux
Grus	The Crane	X	
Hercules	Hercules	*	
Horologium	The Clock	X	
Hydra	The Sea-Serpent	*	
Hydrus	The Water-Snake	X	
Indus	The Indian	X	
Lacerta	The Lizard	–	
Leo	The Lion	*Z	Regulus
Leo Minor	The Little Lion	–	
Lepus	The Hare	–	
Libra	The Scales	*Z	
Lupus	The Wolf	X	
Lynx	The Lynx	–	
Lyra	The Harp	*	Vega
Mensa	The Table	X	
Microscopium	The Microscope	X	
Monoceros	The Unicorn	–	
Musca Australis	The Southern Fly	X	
Norma	The Rule	X	
Octans	The Octant	X	
Ophiuchus	The Serpent-Bearer	*	
Orion	Orion	*	Rigel, Betelgeux
Pavo	The Peacock	X	

Constellations	English Name	Remarks	1st mag. Star or Stars
Pegasus	The Flying Horse	*	
Perseus	Perseus	–	
Phœnix	The Phœnix	X	
Pictor	The Painter	X	
Pisces	The Fishes	*Z	
Piscis Austrinus	The Southern Fish	*	Fomalhaut
Reticulum	The Net	X	
Sagitta	The Arrow	*	
Sagittarius	The Archer	*Z	
Scorpio	The Scorpion	*Z	Antares
Sculptor	The Sculptor	–	
Scutum	The Shield	–	
Serpens	The Serpent	*	
Sextans	The Sextant	–	
Taurus	The Bull	*Z	Aldebaran
Telescopium	The Telescope	X	
Triangulum	The Triangle	*	
Triangulum Australe	The Southern Triangle	X	
Tucana	The Toucan	X	
Ursa Major	The Great Bear	*	
Ursa Minor	The Little Bear	*	
Virgo	The Virgin	*Z	Spica
Volans	The Flying-Fish	X	
Vulpecula	The Fox	–	

Argo Navis is divided up into Carina (the Keel), Vela (the Sails), Puppis (the Poop) and Pyxis Nautica (the Mariner's Compass), though Pyxis is sometimes regarded as an entirely separate constellation.

Some of the original names have been abbreviated; for instance, "Reticulum Rhomboidalis" (the Rhomboidal Net) is simply "Reticulum". A few constellations have alternative names; Scorpio is called "Scorpius" in the list published by the International Astronomical Union, while Ophiuchus may be called "Serpentarius".

PROPER NAMES OF STARS

SOME OF THE stars have been given proper names. Most of these have now fallen into disuse, but since they are still produced occasionally the observer may find it useful to have a list. The names listed here are by no means all that have been given, but includes the more important examples.

A few stars have more than one name (Eta Ursæ Majoris can be "Benetnasch" as well as "Alkaid"), and some names can be spelled in more than one way (Betelgeux can be "Betelgeuse" or "Betelgeuze"). It is clearly pointless to give all these variations.

Constellation	Greek Letter	Name
Andromeda	Alpha	Alpheratz
	Beta	Mirach
	Gamma	Almaak
	Xi	Adhil
Aquarius	Alpha	Sadalmelik
	Beta	Sadalsuud
	Gamma	Sadachiba
	Delta	Scheat
	Epsilon	Albali
Aquila	Alpha	Altair
	Beta	Alshain
	Gamma	Tarazed
	Zeta	Dheneb
	Theta	Ancha
	Kappa	Situla
	Lambda	Althalimain
Ara	Alpha	Choo
Argo Navis	Alpha	Canopus
	Beta	Miaplacidus
	Delta	Koo She
	Epsilon	Avior
	Zeta	Suhail Hadar
	Iota	Tureis
	Kappa	Markeb
	Lambda	Al Suhail Al Wazn

Constellation	Greek Letter	Name
Argo Navis	Xi	Asmidiske
	Rho	Turais
Aries	Alpha	Hamal
	Beta	Sheratan
	Gamma	Mesartim
	Delta	Boteïn
Auriga	Alpha	Capella
	Beta	Menkarlina
	Zeta	Sadatoni
Boötes	Alpha	Arcturus
	Beta	Nekkar
	Gamma	Seginus
	Epsilon	Izar
	Eta	Saak
	h	Merga
	Mu	Alkalurops
Cancer	Alpha	Acubens
	Gamma	Asellus Borealis
	Delta	Asellus Australis
	Zeta	Tegmine
Canes Venatici	Alpha	Cor Caroli
	Beta	Chara
	Schj 152	La Superba
Canis Major	Alpha	Sirius
	Beta	Mirzam
	Gamma	Muliphen
	Delta	Wezea
	Epsilon	Adara
	Zeta	Phurad
	Eta	Aludra
Canis Minor	Alpha	Procyon
	Beta	Gomeisa
Capricornus	Alpha	Al Giedi
	Beta	Dabih
	Gamma	Nashira
	Delta	Deneb al Giedi
	Nu	Alshat
Cassiopeia	Alpha	Shedir
	Beta	Chaph
	Gamma	Tsih
	Delta	Ruchbah
	Theta	Marfak
Centaurus	Alpha	Al Rijil*

* The proper name for Alpha Centauri is not generally used, except by navigators, who refer to it as "Rigel Kent".

Constellation	Greek Letter	Name
Centaurus	Beta	Agena
	Gamma	Menkent
Cepheus	Alpha	Alderamin
	Beta	Alphirk
	Gamma	Alrai
	Xi	Kurdah
Cetus	Alpha	Menkar
	Beta	Diphda
	Gamma	Alkaffaljidhina
	Zeta	Baten Kaitos
	Iota	Deneb Kaitos Shemali
	Omicron	Mira
Columba	Alpha	Phakt
	Beta	Wezn
Corona Borealis	Alpha	Alphekka
	Beta	Nusakan
Corvus	Alpha	Alkhiba
	Gamma	Minkar
	Delta	Algorel
Crater	Alpha	Alkes
Crux Australis	Alpha	Acrux
	Beta	Mimosa*
Cygnus	Alpha	Deneb
	Beta	Albireo
	Gamma	Sadr
	Epsilon	Gienah
	Pi	Azelfafage
Delphinus	Alpha	Svalocin
	Beta	Rotanev
Draco	Alpha	Thuban
	Beta	Alwaid
	Gamma	Etamin
	Delta	Taïs
	Zeta	Aldhibah
	Eta	Aldhibain
	Iota	Edasich
	Lambda	Giansar
	Mu	Alrakis
	Xi	Juza
Equuleus	Alpha	Kitalpha
Eridanus	Alpha	Achernar
	Beta	Kursa

* As in the case of Alpha Centauri, the proper name for Beta Crucis seems to be regarded as "unofficial", and is not generally used.

Constellation	Greek Letter	Name
Eridanus	Gamma	Zaurak
	Delta	Theemini
	Eta	Azha
	Theta	Acamar
	Omicron¹	Beid
	Omicron²	Keid
	Tau	Angetenar
Gemini	Alpha	Castor
	Beta	Pollux
	Gamma	Alhena
	Delta	Wasat
	Epsilon	Mebsuta
	Zeta	Mekbuda
	Eta	Propus
	Mu	Tejat
	Xi	Alzirr
Grus	Alpha	Alnair
	Beta	Al Dhanab
Hercules	Alpha	Rasalgethi
	Beta	Kornephoros
	Zeta	Rutilicus
	Kappa	Marsik
	Lambda	Masym
	Omega	Cujam
Hydra	Alpha	Alphard
Leo	Alpha	Regulus
	Beta	Denebola
	Gamma	Algieba
	Delta	Zosma
	Epsilon	Asad Australis
	Zeta	Adhafera
	Theta	Chort
	Lambda	Alterf
	Mu	Rassalas
Leo Minor	46	Præcipua
Lepus	Alpha	Arneb
	Beta	Nihal
Libra	Alpha	Zubenelgenubi
	Beta	Zubenelchemali
	Gamma	Zubenelhakrabi
	Sigma	Zubenalgubi
Lupus	Alpha	Men
	Beta	Ke Kouan
Lyra	Alpha	Vega

Constellation	Greek Letter	Name
Lyra	Beta	Sheliak
	Gamma	Sulaphat
	Eta	Aladfar
Ophiuchus	Alpha	Rasalhague
	Beta	Cheleb
	Delta	Yed Prior
	Epsilon	Yed Post
	Zeta	Han
	Eta	Sabik
Orion	Alpha	Betelgeux
	Beta	Rigel
	Gamma	Bellatrix
	Delta	Mintaka
	Epsilon	Alnilam
	Zeta	Alnitak
	Eta	Algjebbah
	Kappa	Saiph
	Lambda	Heka
	Upsilon	Thabit
Pegasus	Alpha	Markab
	Beta	Scheat
	Gamma	Algenib
	Epsilon	Enif
Pegasus	Zeta	Homan
	Eta	Matar
	Theta	Biham
	Mu	Sadalbari
	Alpha	Mirphak
	Beta	Algol
	Zeta	Atik
	Xi	Menkib
	Omicron	Ati
	Upsilon	Nembus
Phœnix	Alpha	Ankaa
Pisces	Alpha	Kaïtain
Piscis Australis	Alpha	Fomalhaut
Sagittarius	Alpha	Rukbat
	Beta	Arkab
	Gamma	Alnasr
	Delta	Kaus Meridionalis
	Epsilon	Kaus Australis
	Zeta	Ascella
	Lambda	Kaus Borealis
	Pi	Albaldah

Constellation	Greek Letter	Name
Sagittarius	Sigma	Nunki
Scorpio	Alpha	Antares
	Beta	Graffias
	Gamma (=Sigma Libræ)	Zubenalgubi
	Delta	Dschubba
	Epsilon	Wei
	Theta	Sargas
	Kappa	Girtab
	Lambda	Shaula
	Nu	Jabbah
	Sigma	Alniyat
	Upsilon	Lesath
Serpens	Alpha	Unukalhai
	Theta	Alya
Taurus	Alpha	Aldebaran
	Beta	Alnath
	Gamma	Hyadum Primus
	Epsilon	Ain
	Eta	Alcyone
	17	Electra
	19	Taygete
	20	Maia
	21	Asterope
	23	Merope
	27	Atlas
	28	Pleione
Triangulum	Alpha	Rasalmothallah
Triangulum Australe	Alpha	Atria
Ursa Major	Alpha	Dubhe
	Beta	Merak
	Gamma	Phad
	Delta	Megrez
	Epsilon	Alioth
	Zeta	Mizar
	Eta	Alkaid
	Iota	Talita
	Lambda	Tania Borealis
	Mu	Tania Australis
	Nu	Alula Borealis
	Xi	Alula Australis
	Omicron	Muscida
	Pi	Ta Tsun
	80	Alcor

Constellation	Greek Letter	Name
Ursa Minor	Alpha	Polaris
	Beta	Kocab
	Gamma	Pherkad Major
	Delta	Yildun
	Zeta	Alifa
	Eta	Alasco
Virgo	Alpha	Spica
	Beta	Zawijah
	Gamma	Postvarta
	Epsilon	Vindemiatrix
	Eta	Zaniah
	Iota	Syrma

STARS OF THE "FIRST MAGNITUDE"

THESE ARE THE stars generally regarded as being of the first magnitude. Where the star is a binary system, as with Alpha Centauri, the actual naked-eye magnitude is given, with the spectrum and luminosity of the brighter component. The three apparently brightest single stars are, therefore, Sirius, Canopus and Arcturus. The values for the apparent magnitudes are based on the most recent determinations, and differ in some cases from those previously adopted; in earlier books, for instance, Arcturus was given as +0·24 instead of —0·06. The distances and luminosities of some of the more distant stars, such as Deneb, are naturally rather uncertain.

Star	*Proper name*	*Mag.*	*Spectrum*	*Dist. in Light-yrs.*	*Luminosity Sun = 1*
Alpha Canis Majoris	Sirius	—1·43	A1	8·6	26
Alpha Argûs	Canopus	—0·73	F0	650	80,000
Alpha Centauri	—	—0·27	G2	4	1·1
Alpha Boötis	Arcturus	—0·06	K2	41	100
Alpha Lyræ	Vega	0·04	A0	26	50
Alpha Aurigæ	Capella	0·05	G8	47	150
Beta Orionis	Rigel	0·15	B8	900?	50,000?
Alpha Canis Minoris	Procyon	0·37	F5	10	5
Alpha Eridani	Achernar	0·53	B5	66	200
Alpha Orionis	Betelgeux	var.	M2	190	1,200
Beta Centauri	Agena	0·66	B0	300	3,000
Alpha Aquilæ	Altair	0·80	A7	16	9
Alpha Tauri	Aldebaran	0·85	K5	57	90
Alpha Crucis	Acrux	0·87	B0	230	1,000
Alpha Scorpionis	Antares	0·98	M1	360	3,400
Alpha Virginis	Spica	1·00	B1	230	1,500
Alpha Piscis Australis	Fomalhaut	1·16	A3	24	13
Beta Geminorum	Pollux	1·16	K0	32	28
Alpha Cygni	Deneb	1·26	A2	650	10,000
Beta Crucis	—	1·31	B0	200	850
Alpha Leonis	Regulus	1·36	B7	56	70

STANDARD STARS FOR EACH MAGNITUDE

IT MAY BE helpful to learn the magnitudes of a few standard stars for each magnitude, and the following are suitable. The first-magnitude stars are listed separately, from Sirius (−1·44) to Regulus (+1·36), though Regulus is, of course, nearer 1½ than 1.

Approx. magnitude	Star	Exact magnitude
1½	Epsilon Canis Marjoris	1·48
	Alpha Geminorum (Castor)	1·60
	Lambda Scorpii	1·60
	Gamma Orionis	1·64
2	Alpha Arietis	2·00
	Beta Ursæ Minoris (Kocab)	2·04
	Kappa Orionis	2·06
	Alpha Andromedæ (Alpheratz)	2·06
2½	Gamma Ursæ Majoris (Phad)	2·44
	Epsilon Cygni	2·46
	Alpha Pegasi	2·50
	Delta Leonis	2·57
3	Zeta Aquilæ	2·99
	Gamma Boötis	3·05
	Delta Draconis	3·06
	Zeta Tauri	3·07
3½	Alpha Trianguli	3·45
	Zeta Leonis	3·46
	Beta Boötis	3·48
	Epsilon Tauri	3·54
4	Beta Aquilæ	3·90
	Gamma Coronæ Borealis	3·93
	Delta Ceti	4·04
	Delta Cancri	4·17
4½	Nu Andromedæ	4·42
	Delta Ursæ Minoris	4·44
	Nu Cephei	4·46
	Psi Ursæ Majoris	4·54

Approx. magnitude	Star	Exact magnitude
5	Rho Ursæ Majoris	4·99
	Eta Ursæ Minoris	5·04
	Delta Trianguli	5·07
	Zeta Canis Minoris	5·11
5½	Theta Ursæ Minoris	5·33
	Rho Coronæ Borealis	5·43
	Epsilon Trianguli	5·44

Appendix XIX

THE GREEK ALPHABET

α	Alpha	ν	Nu
β	Beta	ξ	Xi
γ	Gamma	ο	Omicron
δ	Delta	π	Pi
ε	Epsilon	ρ	Rho
ζ	Zeta	σ	Sigma
η	Eta	τ	Tau
θ	Theta	υ	Upsilon
ι	Iota	φ	Phi
κ	Kappa	χ	Chi
λ	Lambda	ψ	Psi
μ	Mu	ω	Omega

STELLAR SPECTRA

Type	Surface Temp., degrees C	Colour	Typical Star	Remarks
W	36,000+	Greenish white	Gamma Argûs, WC7	Wolf-Rayet. Many bright lines; helium prominent
O	36,000+	Greenish white	Zeta Argûs, O5	Wolf-Rayet. Helium prominent
B	28,600	Bluish	Spica, B1	Helium prominent
A	10,700	White	Sirius, A1	Hydrogen lines prominent
F	7,500	Yellowish	Beta Cassiopeiæ, F2	Calcium lines prominent
G (giant)	5,200	Yellow	Epsilon Leonis, G0	Metallic lines very numerous
G (dwarf)	6,000	Yellow	Sun, G2	
K (giant)	4,230	Orange	Arcturus, K2	Hydrocarbon bands appear
K (dwarf)	4,910	Orange	Epsilon Eridani, K2	
M (giant)	3,400	Orange-red	Betelgeux, M2	Broad titanium oxide and calcium bands or flutings
M (dwarf)	3,400	Orange-red	Wolf 359, M6	
R	2,300	Orange-red	U Cygni	Carbon bands
N	2,600	Red	S Cephei, Nc	Carbon bands. Reddest of all stars
S	2,600	Red	R Andromedæ	Some zirconium oxide bands. Mostly long-period variables

A separate class, Q, is reserved for novæ.

Appendix XXI

LIMITING MAGNITUDES AND SEPARATIONS FOR VARIOUS APERTURES

IT IS EXTREMELY difficult to give definite value for limiting magnitudes and separations, since so much must depend upon individual observers. The following table must be regarded as approximate only. The third column refers to stars of equal brilliancy and of about the sixth magnitude. Where the components are unequal, the double will naturally be a more difficult object, particularly if one star is much brighter than the other.

Aperture of O.G. in inches	Faintest magnitude	Smallest separation, seconds of arc
2	10·5	2·5
3	11·4	1·8
4	12·0	1·3
5	12·5	1·0
6	12·9	0·8
7	13·2	0·7
8	13·5	0·6
10	14·0	0·5
12	14·4	0·4
15	14·9	0·3

Appendix XXII

ANGULAR MEASURE

IT MAY BE useful to give the angular distances between some selected stars, as this will be of use to those who are not used to angular measurement. The distance all round the horizon is of course 360 degrees, and from the zenith to the horizon 90 degrees; the Sun and Moon have angular diameters of about 0·5 degrees, which is the same as that of a halfpenny held at a distance of 9½ feet from the eye.

Degrees (approx.)	Stars
60	Polaris to Pollux: Alpha Ursæ Majoris to Beta Cassiopeiæ.
50	Sirius to Castor: Polaris to Vega
45	Polaris to Deneb: Spica to Antares
40	Capella to Betelgeux: Castor to Regulus
35	Vega to Altair: Capella to Pollux
30	Polaris to Beta Cassiopeiæ: Aldebaran to Capella
25	Sirius to Procyon: Vega to Deneb
20	Betelgeux to Rigel: Procyon to Pollux
15	Alpha Andromedæ to Beta Andromedæ: Alpha Centauri to Acrux
10	Betelgeux to Delta Orionis: Acrux to Agena
5	Alpha Ursæ Majoris to Beta Ursæ Majoris
4½	Castor to Pollux: Alpha Centauri to Agena
3	Beta Scorpionis to Delta Scorpionis
2½	Altair to Beta Aquilæ
2	Altair to Gamma Aquilæ: Beta Lyræ to Gamma Lyræ
1½	Beta Arietis to Gamma Arietis
1	Atlas to Electra (Pleiades)

To find the diameter of a telescopic field, select some star very near the celestial equator (such as Delta Orionis or Zeta Virginis) and allow it to drift through the field. This time in minutes and seconds, multiplied by 15, will give the angular diameter of the field in minutes and seconds of arc. For instance, if Delta Orionis takes 1 minute 3 seconds to pass, through the field, the diameter is 1 minute 3 seconds × 15, or 15′ 45″ of arc.

Appendix XXIII

TEST DOUBLE STARS

THE FOLLOWING list is only approximate, since again so much depends upon the observer as well as upon the precise conditions, but it may be useful as a rough guide.

Apertures of O.G., inches	Star	Mags.		Separation, secs. of arc	Position Angle, deg.
I	Acrux	1·4,	1·9	4·7	119
	Alpha Herculis	var.,	5·4	4·5	109
2	Rigel	0·1,	6·7	9·5	202
	Gamma Leonis	2·3,	3·5	4·3	122
	Epsilon Boötis	2·4,	5·0	2·8	334
3	Polaris	2·0,	8·9	18·3	217
	Theta Virginis	4·0,	9·0	7·2	343
4	Theta Aurigæ	2·7,	7·1	2·8	332
	Eta Orionis	3·8,	4·8	1·4	079
	Delta Cygni	2·9,	6·4	2·1	240
	Iota Ursæ Majoris	3·1,	10·8	7·4	002
5	Zeta Boötis	4·6,	4·6	1·2	309
	Omega Leonis	5·9,	6·7	1·0	129
8	Lambda Cassiopeiæ	5·5,	5·8	0·6	180
	*Gamma² Andromedæ	5·4,	6·6	0·7	109
9	Eta Coronæ Borealis	5·7,	5·9	0·7	069

* Gamma² Andromedæ is the smaller component of the easy double Gamma Andromedæ.

Appendix XXIV

EXTINCTION

WHEN ESTIMATING the brightness of a naked-eye variable, care must be taken to allow for atmospheric dimming. The closer a star is to the horizon, the more of its light will be lost. The following table gives the amount of dimming for various altitudes above the horizon. Above an altitude of 45 degrees, extinction can be neglected for all practical purposes.

Altitude degrees	Dimming in magnitudes
1	3
2	2·5
4	2
10	1
13	0·8
15	0·7
17	0·6
21	0·4
26	0·3
32	0·2
43	0·1

NAKED-EYE NOVÆ

THIS LIST includes all novæ since 1572 that have become bright enough to be visible with the naked eye. An asterisk denotes that the nova was too far south to be visible in New York.

Year	Nova	Maximum Mag.	Discoverer
1572	Cassiopeiæ	−4	Tycho Brahe
1600	P Cygni	3	Blaeu
1604	Ophiuchi	−2·3	(?)
1670	Vulpeculæ	3	Anthelm
1783	Sagittæ	6	D'Agelet
1848	Ophiuchi	4	Hind
1866	T Coronæ	2	Birmingham
1876	Cygni	3	Schmidt
1891	Aurigæ	4·2	Anderson
1898	Sagittarii	4·9	Miss Fleming
1901	Persei	0·0	Anderson
1903	Geminorum	5·0	Turner
1910	*Aræ	6·0	Miss Fleming
1910	Lacertæ	4·6	Espin
1912	Geminorum	3·3	Enebo
1918	Aquilæ	−1·1	Bower
1918	Monocerotis	5·7	Wolf
1920	Cygni	2·0	Denning
1925	*RR Pictoris	1·1	Watson
1927	Tauri	6·0	Schwassmann, Wachmann
1934	DQ Herculis	1·2	Prentice
1936	Aquilæ	5·4	Tamm
1936	Lacertæ	1·9	Gomi
1936	Sagittarii	4·5	Okabayasi
1939	Monocerotis	4·3	Whipple, Wachmann
1942	Argûs	0·4	Dawson
1950	Lacertæ	6·0	Bertaud
1960	Herculis	5·0	Hassell
1963	Herculis	3·2	Dahlgren, Peltier
1967	Delphini	4·5	Alcock

Nova Argûs (Puppis) 1942 was extremely difficult to observe in England owing to its low altitude.

MESSIER'S CATALOGUE

MESSIER'S FAMOUS catalogue of nebular objects includes most of the brightest nebulæ and clusters visible in England. It is therefore useful to give his list, as most of the objects can be found by means of the star maps in Appendix XXVI and can be picked up by means of small telescopes.

Number	Constellation	Type	Magnitude	Remarks
1	Taurus	Wreck of super-nova	8·4	Crab Nebula. Radio scurce
2	Aquarius	Globular	6·3	
3	Canes Venatici	Globular	6·4	
4	Scorpio	Globular	6·4	
5	Serpens	Globular	6·2	
6	Scorpio	Open cluster	5·3	
7	Scorpio	Open cluster	4·0	
8	Sagittarius	Nebula	6·0	Lagoon Nebula
9	Ophiuchus	Globular	7·3	
10	Ophiuchus	Globular	6·7	
11	Scutum	Open cluster	6·3	Wild Duck Cluster
12	Ophiuchus	Globular	6·6	
13	Hercules	Globular	5·7	Great globular cluster
14	Ophiuchus	Globular	7·7	
15	Pegasus	Globular	6·0	
16	Serpens	Nebula and embedded cluster	6·4	
17	Sagittarius	Nebula	7·0	Omega· or Horseshoe (Nebula
18	Sagittarius	Open cluster	7·5	
19	Ophiuchus	Globular	6·6	
20	Sagittarius	Nebula	9·0	Trifid Nebula
21	Sagittarius	Open cluster	6·5	

Number	Constellation	Type	Magnitude	Remarks
22	Sagittarius	Globular	5·9	
23	Sagittarius	Open cluster	6·9	
24	Sagittarius	Open cluster	4·6	
25	Sagittarius	Open cluster	6·5	
26	Scutum	Open cluster	9·3	
27	Vulpecula	Planetary	7·6	Dumbbell Nebula
28	Sagittarius	Globular	7·3	
29	Cygnus	Open cluster	7·1	
30	Capricornus	Globular	8·4	
31	Andromeda	Spiral galaxy	4·8	Great Galaxy
32	Andromeda	Elliptical galaxy	8·7	Satellite of M.31
33	Triangulum	Spiral galaxy	6·7	Triangulum Spiral
34	Perseus	Open cluster	5·5	
35	Gemini	Open cluster	5·3	
36	Auriga	Open cluster	6·3	
37	Auriga	Open cluster	6·2	
38	Auriga	Open cluster	7·4	
39	Cygnus	Open cluster	5·2	
41	Canis Major	Open cluster	4·6	
42	Orion	Nebula	4±	Great Nebula in Orion
43	Orion	Nebula	9±	Part of Orion Nebula
44	Cancer	Open cluster	3·7	Præsepe
45	Taurus	Open cluster	—	Pleiades
46	Argo Navis	Open cluster	6·0	In Puppis
49	Virgo	Elliptical galaxy	8·6	
50	Monoceros	Open cluster	6·3	
51	Canes Venatici	Spiral galaxy	8·1	Whirlpool Galaxy Radio source
52	Cassiopeia	Open cluster	7·3	
53	Coma Berenices	Globular	7·6	
54	Sagittarius	Globular	7·3	
55	Sagittarius	Globular	7·6	
56	Lyra	Globular	8·2	
57	Lyra	Planetary	9·3	Ring Nebula
58	Virgo	Spiral galaxy	8·2	
59	Virgo	Elliptical galaxy	9·3	
60	Virgo	Elliptical galaxy	9·2	

261

Number	Constellation	Type	Magnitude	Remarks
100	Coma Bernices	Spiral galaxy	10·6	
101	Ursa Major	Spiral galaxy	9·6	
103	Cassiopeia	Open cluster	7·4	
104	Virgo	Spiral galaxy	8·4	"Sombrero Hat" Galaxy

Various forms of the Messier catalogue have been given, notably by Owen Gingerich (*Sky and Telescope*, Vol. XIII, p. 158 (1954)) and R. H. Garstang (*B.A.A. Handbook*, 1964; page 63). Five additions were made, all objects observed by the French astronomer Méchain, and these are often included in the catalogue: M. 105 (elliptical galaxy in Leo), M. 106 (spiral galaxy in Canes Venatici), M. 107 (globular in Ophiuchus), and M. 108 and 109 (spiral galaxies in Ursa Major).

M. 40 is not identifiable; it may be simply a couple of faint stars, or it may have been a comet. M. 91 is also an absentee, and this too may have been a comet, though Gingerich suggests that it may be identical with M. 58. There is grave doubt about the identities of M. 47 and M. 48; it has been suggested that M. 47 is an open cluster in Argo Navis (Puppis) and M. 48 an open cluster in Hydra. M. 102 may have been identical with M. 101, or it may possibly have been a faint spiral galaxy in Draco. Finally, M. 73 consists of four faint, unconnected stars, and is not a true cluster or nebular object.

THE STAR MAPS

MANY PERIODICALS and some of the national newspapers give regular "stars of the month" charts. These are useful, but in my personal opinion they are of limited help to the absolute beginner, since they show so many objects that confusion is bound to result.

I have found that the best way to learn the various groups is to pick them out, one by one, by means of the two leading constellations of our skies, Ursa Major (the Great Bear) and Orion. Of these, Orion is the more brilliant, but it is not always visible in New York, whereas the Bear never sets.

Using these two constellations as "signposts in the sky", it is possible to identify the other groups, and this system is developed in the maps given here. The key maps, I and II, will enable the beginner to find his way about in Maps IV to X. There can be little difficulty in finding Orion and the Great Bear; for one thing, there will always be someone near by who knows them.

The star-maps given here are not precision charts; nor are they intended to be, but it is hoped that they will be of some use as an aid to finding one's way about the sky.

In the constellation notes, all stars down to magnitude 3·5 have been listed under the heading "Chief Stars". All the doubles, variables and clusters mentioned are easy objects.

MAP I. KEY MAP: URSA MAJOR (THE GREAT BEAR)

Everyone knows the Great Bear or Dipper. Its seven stars are a familiar feature of the night sky, and it is of course so far north that it never sets in the latitude of New York. The proper names of the seven are frequently used: in addition, Merak and Dubhe are popularly known as the "Pointers".

The first step after having identified the Bear is to find the Pole Star. Imagine a line drawn from Merak through Dubhe, and prolonged; it will reach a second-magnitude star rather "out on its own", and this is Polaris. The Little Bear, Ursa Minor, can then be picked out, bending back towards the Great Bear itself. The stars are much fainter, but one of them, the rather reddish Kocab, is of magnitude 2.

Now imagine a line from Alioth, in the Great Bear, through Polaris. Prolonged for an equal distance on the far side of Polaris, it will reach five brightish stars (magnitudes 2 to 3) arranged in a rough W. This is Cassiopeia, which, like the Bears, never sets in England.

A line from Megrez through Dubhe will come eventually to Capella, which is one of the brightest stars in the entire sky. It is circumpolar in England, but at its lowest, as during summer evenings, it almost reaches the horizon. In winter evenings it is high up, and may indeed pass overhead. If you see a really bright star straight above you, it can be only Capella or Vega; Capella is yellowish, and may be recognized by the small triangle of stars close by it, whereas Vega is decidedly blue. Vega can be found by means of a line beginning at Phad, passing between Megrez and Alioth, and prolonged for some distance across the sky.

The remaining stars shown in Map I are not circumpolar. The Twins, Castor and Pollux, may be found by means of a line from Megrez through Merak; they are at their best in winter. Regulus and the other stars of the Lion, found by a line from Megrez through Phad, seem to follow the Twins in the sky; the curved arrangement of stars rather like a reversed question-mark, of which Regulus is the brightest, is known as the "Sickle of Leo", and is easy to recognize. Even easier is Arcturus, about as bright as Capella and Vega. This is found by means of a line from Mizar through Alkaid, and curved

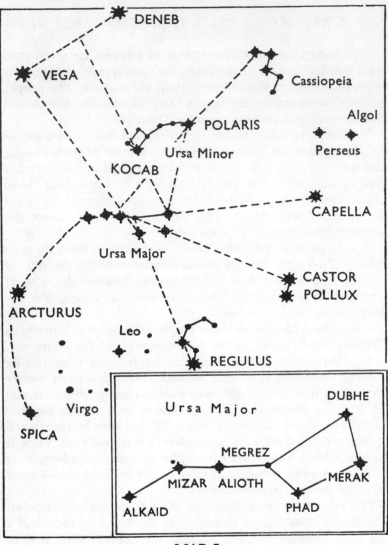

MAP I

rather downwards; if the curve is continued through Arcturus it comes to another 1st-magnitude star, Spica in Virgo. Arcturus and Spica are prominent features of the spring and summer skies of England.

It may be added that Arcturus shines with a distinctly orange light, so that it cannot be confused with Capella or Vega.

MAP II. KEY MAP: ORION

It is a pity that Orion is not circumpolar in America, as it is a magnificent "signpost", as well as being a beautiful constellation in itself. It cannot be mistaken, as all its chief stars are brilliant, two of the first magnitude (Betelgeux and Rigel) and five of the second. Mintaka, Alnilam and Alnitak form the famous Belt. The periods of visibility of Orion in England can be judged from the following:

January 1st   Rises 4 p.m., highest 10 p.m., sets 5 a.m.
April 1st   Rises in daylight, highest in daylight, sets 11 p.m.
July 1st   Rises 4 a.m., highest in daylight, sets in daylight.
October 1st   Rises 10 p.m., highest 5 a.m., sets in daylight.

It must be understood that these times are only very rough; Orion covers a considerable area, and takes some time to "rise". It is however clear that the constellation is best seen in winter and in the early mornings in autumn.

The first-magnitude stars in the key map are easy to find if Orion can be seen. The three stars of the Belt (Mintaka, Alnilam and Alnitak) point downwards to Sirius, which is the most brilliant star in the sky, though of course less bright than Venus, Jupiter or Mars when well placed. Upwards, the Belt stars indicate Aldebaran in Taurus, a reddish first-magnitude star of about the same colour and brightness as Betelgeux.

Bellatrix and Betelgeux point more or less to Procyon, in Canis Minor, which is not much fainter than Rigel; if this line is continued and curved slightly it reaches a reddish 2nd-magnitude star, Alphard in Hydra, known as "the Solitary One" because it lies in a very barren region. The Twins, Castor and Pollux, can be found by a line from Rigel through Betelgeux; since they can also be found by using Ursa Major, this links the two key maps. Capella is indicated by a line from Saiph through Alnitak. Diphda in Cetus, the other star shown in the diagram, is less easy to find. It is only of mag. 2, and is frequently visible when Orion is below the horizon.

Undoubtedly a winter evening is the best time to start star recognition, since then both our "signposts", Orion and the Bear, can be seen. If a start be made in summer, we must do without Orion; but the Bear can by itself teach us the way

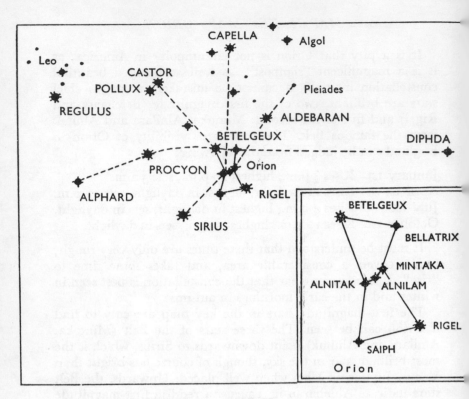

MAP II

about the heavens, and even though the stars seem at first to be arranged in a chaotic manner it takes surprisingly little time to find one's way about.

Each of the following charts contain at least one key map object. Exact positions of telescopic objects, in right ascension and declination, are not given here, because an observer who possesses a telescope equipped with setting circles will in any case need a more detailed set of charts. By far the best star atlas for the average worker is *Norton's*, published by Gall and Inglis.

MAP III. URSA MAJOR, URSA MINOR, DRACO, CEPHEUS, CAMELOPARDUS

This is the North Polar region. The stars in it are of course circumpolar in New York, and will quickly be recognized.

URSA MAJOR. This has already been described at length. The chief stars are Epsilon (Alioth) and Alpha (Dubhe) (1·8), Eta (Alkaid) (1·9), Zeta (Mizar) (2·1), Beta (Merak) and Gamma (Phad) (2·4), Psi and Mu (3·0), Iota (3·1), Theta (3·2), Delta (Megrez) (3·3) and Lambda (3·4). Part of the constellation extends on to Map VI.

Double Star. Zeta (Mizar). Naked-eye pair with Alcor. In a low power Mizar is itself double; mags. 2·2, 3·9; dist. 14″·5; P.A. 150°. Between this pair and Alcor is another star. Mizar is one of the finest doubles for a small aperture.

Variables. T: mag. 5·5 to 13, period 254 days. Spectrum Me: red. An easy object near maximum.

R: mag. 5·9 to 13; period 298 days. Spectrum Me: red. Like T, an easy object near maximum.

Clusters and Nebulæ. M.81 and M.82; two galaxies, close together, identifiable without much difficulty.

M.97: The Owl Nebula, a planetary so called because its two hot stars do give it the look of an owl's face with high powers. It is very faint with small apertures, but is worth looking for.

URSA MINOR curves down over the stars of Ursa Major. Chief stars: Alpha (Polaris) (2·0), Beta (Kocab) (2·0), Gamma (3·1). Kocab is a fine orange star.

Double Star. Polaris. Mags. 2·0, 9·0; distance 18″·3, P.A. 217°. An easy object with aperture 3 in. or more.

DRACO. A long, winding constellation, stretching from Lambda (between Dubhe and Polaris) as far as Gamma, which lies near Vega. The chief stars are Gamma (2·2), Eta (2·7), Beta (2·8), Delta (3·1), Zeta (3·2) and Iota (3·3). Alpha or Thuban (3·6) used to be the pole star in ancient times.

Double Stars. Nu; magnitudes 4·5, 4·5; distance 62″. This is a very wide, easy double.

Eta: magnitudes 2·7, 8·0; distance 6″; P.A. 142°. This can be seen with a 3-inch refractor.

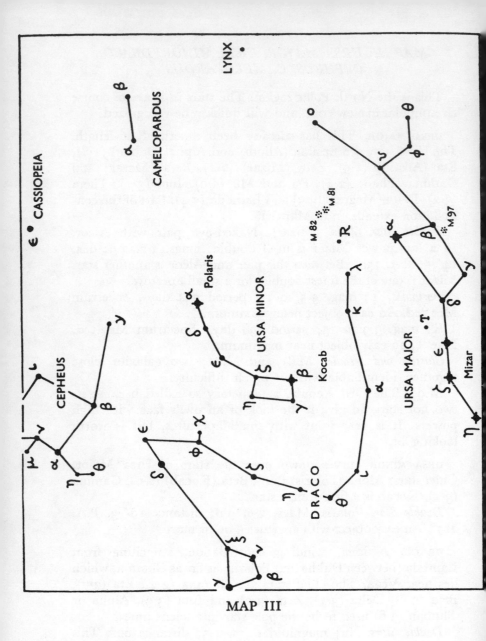

MAP III

Epsilon: magnitudes 4·0, 7·5; distance 3″·3; P.A. 009°.

CEPHEUS is not one of the easier constellations to identify, but it is useful to remember that Gamma Cephei lies more or less between Polaris and the W. of Cassiopeia. It is better shown on Map VII. Chief stars: Alpha (2·4), Beta (3·1), and Gamma (3·2). Telescopic objects are given in the notes on Map VII.

CAMELOPARDUS. A large, dull constellation, with no stars brighter than the 4th magnitude, and with no objects of particular interest. It is in fact one of the most barren regions of the heavens.

MAP IV. ORION, LEPUS, ERIDANUS, TAURUS, CETUS, AURIGA, COLUMBA, CÆLUM, FORNAX

The times of rising and setting of Orion, in England, were given in the notes on Map II. Capella is just circumpolar, but can almost graze the horizon. Perseus is shown in part, and also Triangulum.

ORION is probably the most glorious constellation in the heavens, and is easy to recognize. Betelgeux is a fine sight with a lower power (spectrum M; orange-red), while Rigel is brilliantly white. Rigel appears only very slightly less brilliant than Arcturus and Vega. The other leading stars are Gamma (Bellatrix) (1·6), Epsilon (Alnilam) (1·7), Zeta (Alnitak) (1·8), Kappa (Saiph) (2·1), Delta (Mintaka) (2·3, but slightly variable), Iota (2·8), Pi³ (3·2), Eta (3·4) and Lambda (3·5).

Double Stars. Rigel: magnitudes 0·1, 6·7; distance 9″·4; P.A. 202°. A test for a 2-in. O.G.; easy with a 3-in. The companion is said to be bluish, but to me it always appears white.

Eta; magnitudes 3·6, 4·8; distance 1″·4; P.A. 080°.

Lambda: magnitudes 3·6, 5·5; distance 4″·2; P.A. 043°.

Zeta: magnitudes 1·9, 5·0; distance 2″·8; P.A. 162°. I find this very hard with anything less than 3-in. aperture.

Iota: magnitudes 3·0, 7·4; distance 11″·4; P.A. 140°. Immersed in nebulosity.

Theta: the Trapezium, a multiple star. Magnitudes 6·0,

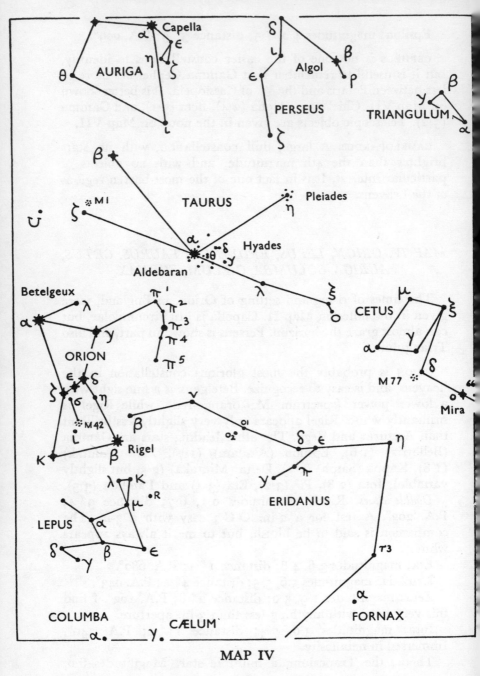

MAP IV

7·0, 7·0, 7·5. All four stars are easy in a 3-in. O.G. Immersed in the Great Nebula, M.42.

Sigma: another multiple. The magnitudes of the four brightest stars are 4·0, 7·0, 7·5 and 9·9. Less striking than Theta, but well worth examination.

Delta: magnitudes 2·3 (var.), 6·7; distance 53″; P.A. 000°. Very wide and easy.

Variables. Betelgeux: 0·0 to 1·2. This is a greater range than is given in most textbooks, but Sir John Herschel recorded that he saw it outshine Rigel, and this has also been my experience. The best comparison star for normal periods is of course Aldebaran; another, useful when Betelgeux is faint, is Pollux.

U (not far from Zeta Tauri). Magnitudes 5·5 to 12·6. Period 375 days. A red, Me-type long-period variable.

Delta (Mintaka): an eclipsing binary of small magnitude range (2·20 to 2·35).

Clusters and Nebulæ. M.42: the Sword of Orion, visible to the naked eye, and the most prominent of all galactic nebulæ. It is a splendid sight in a small telescope; dark nebulosity may be seen close to the Trapezium.

LEPUS is a small constellation near Orion. The chief stars are Alpha (2·6), Beta (2·8), Epsilon (3·2) and Mu (3.3).

Double Stars. Kappa: magnitudes 4·9, 7·5; distance 2″·6; P.A. 000°. The primary is yellowish and the companion bluish.

Beta: magnitudes 2·8, 9·4; distance 2″·5; P.A. 313°.

Variable. R: magnitude 5·9 to 10·5; period 430 days. This is an intensely red star of spectrum N. It is not hard to find when near maximum.

COLUMBA lies below Lepus, and is too far south to be well seen in England. Chief stars: Alpha (2·6), Beta (3·1). The constellation contains no features of particular interest.

CÆLUM has no star brighter than Alpha (4·5), and is always very low in our latitudes.

Double Star: Gamma: magnitudes 4·7, 8·5; distance 3″; P.A. 310°.

ERIDANUS. A very long constellation, of which the chief stars are the first-magnitude Achernar, and Beta (2·8), Theta

(2·9) and Gamma (3·0). Achernar and Theta never rise in New York, but are shown in Maps XI and XII.

Double Star. Omicron²: magnitudes 4·0, 9·0: distance 82″: P.A. 107°. Theta is double; separation 8″.

FORNAX has no star brighter than Alpha (4·0). It is low in England, and contains no features of interest.

CETUS is another long, winding constellation; the rest of it is shown in Map X. Chief stars: Beta (2·0), Alpha (2·5), Eta and Tau (3·5). Alpha is a fine orange star.

Double Stars. Gamma: magnitudes 3·6, 6·2; distance 3″; P.A. 295°.

66: magnitudes 6·0, 7·8; distance 16″·3; P.A. 232°. The primary is yellow and the companion blue. This is in a low-power field with Mira, and is a useful guide when Mira is faint.

Variable. Omicron (Mira): magnitude 1·7 to 9·6; period 331 days. This interesting star is fully described in the text.

Nebula. M.77: a fairly easy object, one degree away from Delta. It is actually a spiral galaxy.

TAURUS. This is a Zodiacal constellation of great interest. Apart from Aldebaran, the chief stars are Beta (1·6), Eta (Alcyone) (2·9), Zeta (3·1), and two of the Hyads, Theta² (3·4) and Epsilon (3·5). Beta Tauri is also known as Gamma Aurigæ.

Double Star. Aldebaran has a 13th magnitude companion; distance 121″; P.A. 034°. This is a wide optical double, but the faintness of the companion makes it a useful test object.

Variable. Lambda: magnitude 3·3 to 4·2; period 3·9 days. Spectrum B3. This is an eclipsing binary of the Algol type.

Clusters and Nebulæ. M.1: the remarkable "Crab Nebula", near Zeta, described in the text.

The Pleiades and Hyades are also described in the text. The Hyades, which are scattered, are best seen in binoculars.

AURIGA. One of the brightest of the northern groups. Capella is shown in both Key Maps, and is surpassed by only three other stars visible from England: Sirius, Vega and Arcturus. The difference between Vega and Capella is only 1/100 of a magnitude. Capella is yellow, and can be identified by the three fainter stars (Epsilon, Zeta, Eta) close by it; these have been termed the "Hædi", or Kids. Gamma Aurigæ is now known as Beta Tauri. This is one of a few cases of stars being included in two constellations; others are Alpha Andromedæ (=Delta Pegasi) and Gamma Scorpionis (=Sigma Libræ).

The other chief stars of Auriga are Beta (1·9), Iota and Theta (2·6) and Eta (3·2). Epsilon, the vast giant, is variable; it is comparable with Eta. The magnitude range is small, and this applies also to the other giant eclipsing binary, Zeta, whose fluctuations will not easily be detected without instruments.

Epsilon's magnitude varies from 3·1 to about 3·7, and the period is just over 27 years.

Double Star. Theta: magnitudes 2·6, 7·1; distance 2"·8; P.A. 333°. I always find this rather difficult with a 6-in. reflector; it is said to be a test for a 4-in. O.G.

MAP V. GEMINI, CANCER, CANIS MAJOR, CANIS MINOR, MONOCEROS, HYDRA

The constellations shown in this map are at their best in winter and spring evenings. The following times of rising and setting in England are for Cancer, and are of course very rough. Cancer is a Zodiacal constellation, as are Gemini and Leo.

January 1st Rises 6 p.m., highest 2 a.m., set in daylight.
April 1st Rises in daylight, highest 8 p.m., sets 4 a.m.
July 1st Rises in daylight, highest in daylight, sets in daylight.
October 1st Rises at midnight, highest 8 a.m., sets in daylight.

CANIS MAJOR is most notable because of the presence of Sirius, the brightest star in the sky. The other chief stars are Epsilon (1·5), Delta (1·8), Beta (2·0), Eta (2·5), and Omicron² and Zeta (3·0). The group is easy to find from Orion. Actually there are few interesting telescopic objects in Canis Major, but M.41 is a bright cluster well worth looking at.

CANIS MINOR contains Procyon; the only other bright star is Beta (2·9).

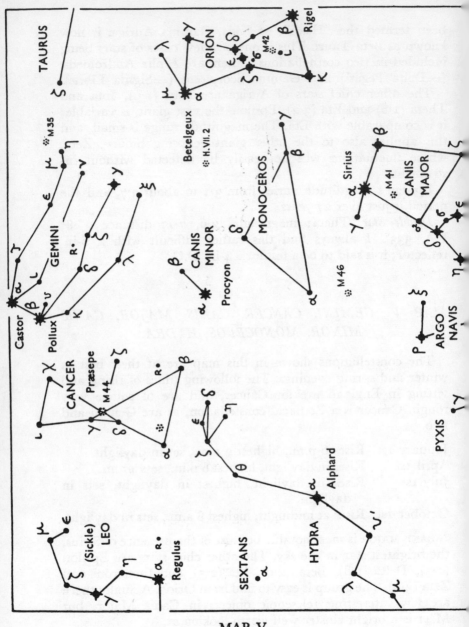

MAP V

MONOCEROS. A large, faint constellation with no star brighter than the fourth magnitude; it lies in the area enclosed by Procyon, Sirius and Betelgeux. The Milky Way passes through is, and there are some rich telescopic fields, so that the region it worth sweeping with low powers.

Double Star. Beta: a triple. Magnitudes 5·0, 5·5, 5·9; distances 7"·4 and 2"·8; P.A.s 132° and 105°.

Cluster. Around the 6th magnitude star 12 Monocerotis is a fine open cluster, H.VII.2 (not in Messier's catalogue). It lies between Betelgeux and the fourth-magnitude star Delta Monocerotis.

ARGO NAVIS. Nearly all of this grand constellation, including Canopus, is too far south to be seen in New York; it is shown in Map XIV. A few stars of Puppis, including Rho (2·7) and Xi (3·3) can be made out, and a few stars of PYXIS NAUTICA can also be seen low down on the horizon.

Cluster. M.46: a beautiful small cluster, roughly between Rho Argûs and Alpha Monocerotis.

GEMINI. This is one of the grandest of all constellations. As well as Pollux and Castor, it includes other bright stars: Gamma (1·9), Mu and Epsilon (3·0), Xi (3·4) and Delta (3·5). Moreover, the Milky Way passes through it. Castor and Pollux can be found from either key map. Pollux, rather orange in colour (type K) is now appreciably brighter than Castor, though it seems that in Ptolemy's day this was not the case. There are plenty of interesting objects in Gemini.

Double Stars. Castor: magnitudes 1·9, 2·8; distance 1·8"; P.A. 151°. A fine double; binary, period 380 years. As described in the text, Castor is a multiple system; Castor C, magnitude 9·1, lies at 73".

Delta: magnitudes 3·5, 8·2; distance 6"·7; P.A. 210°. Test for a 2-in. O.G., though I always find it rather easy with such an aperture.

Lambda: magnitudes 3·7, 10·0; distance 10"; P.A. 033°.

Kappa: magnitudes 4·0, 8·5; distance 6"·7; P.A. 235°.

Variables. Eta: a long-period M-type variable; magnitude 3·3 to 4·2; period 231 days.

Zeta: magnitude 3·7 to 4·3; period 10·2 days. Spectrum G. A typical Cepheid. A useful comparison star is Nu (4·1).

R: magnitude 6·0 to 14; period 370 days.

Cluster. M.35. A fine open cluster, a splendid sight in a small telescope. Mu and Eta act as excellent "guide stars" to it.

CANCER. A faint constellation, the brightest stars being Beta (3·8), Iota (4·1) and Delta (4·2), but it includes some interesting objects, such as Præsepe. It is not unlike a very dim and ghostly Orion, and lies in the area enclosed by Pollux, Procyon and Regulus.

Double Stars. Zeta: magnitudes 5·0, 5·7; distance 1″; binary, period 60 years. It was at its widest in the year 1960, and is now closing up again. There is a third component; magnitude 6·1, distance 5″·7.

Iota: magnitudes 4·3, 6·3; distance 31″; P.A. 307°. The larger star is yellowish, the companion bluish.

Variable. R: magnitude 5·9 to 11·5; period 362 days. An M-type, long-period variable.

Clusters. M.44 (Præsepe). One of the best of the open clusters. It has been described in the text, and can be seen with the naked eye on any reasonably transparent moonless night.

M.67. A conspicuous telescopic object close to Alpha (4·3).

HYDRA. Apart from Argo Navis, which is now divided up, Hydra is the largest constellation in the sky; parts of it are also shown on Maps VI and IX. It is however rather barren. The chief stars are Alpha (2·0), Gamma (3·0), Zeta and Nu (3·1), Pi (3·2) and Epsilon (3·4). Alphard is shown on the Key Map II; it is easy to find, as it is distinctly reddish, and appears very isolated. Its name of "the Solitary One" suits it well. It can be identified by continuing the "sweep" from Bellatrix through Betelgeux and Procyon and, incidentally, Castor and Pollux point to it. It is a fine object in a low power.

Double Stars. Theta: magnitudes 4·9, 10·8; distance 38″; P.A. 185°. The faintness of the companion makes it a useful test.

Epsilon: magnitudes 3·6, 7·7; distance 3″·6; P.A. 253°. This is in the "head" of Hydra, which is easy to find, as it lies roughly midway between Procyon and Regulus. The bright component is a close binary with a period of 15 years. A 12-mag. star lies at 20″.

SEXTANS. A faint and unremarkable constellation, with no star as bright as the 4th magnitude, and no interesting telescopic objects.

Parts of Leo and Taurus are included in this map, but are better shown in Maps VI and IV respectively.

MAP VI. LEO, VIRGO, COMA BERENICES, CORVUS, CRATER, LEO MINOR

This area contains some interesting features; Regulus, Spica and Arcturus are shown in Key Map I. The rough times of rising and setting for Spica, in England, are:

January 1st   Rises 1 a.m., highest 6 a.m., sets in daylight.
April 1st   Rises 7 p.m., highest midnight, sets 5 a.m.
July 1st   Rises in daylight, highest in daylight, sets 11 p.m.
October 1st   Rises in daylight, highest in daylight, sets in daylight.

LEO. A large, important constellation. Regulus is of course the chief star, and other bright stars are Gamma (2·0), Beta (2·1), Delta (2·6), Epsilon (3·0), Theta (3·3) and Zeta (3·5). The curved line of stars beginning with Regulus is known as the Sickle, and is a prominent feature; the triangle formed by Beta, Delta and Theta is also easy to find. Beta is a secular variable. Ptolemy made it of the 1st magnitude, but it is now below the second. As it is suspected of variability, it is well worth watching; Gamma makes a good comparison star.

Double Stars. Gamma: magnitudes 2·3, 3·8; distance 4"·3; P.A. 121°. A fine binary, with a period of 407 years.

Iota: magnitudes 3·9, 7·0; distance 0"·6; P.A. 015°.

Variable Star. R: magnitude 5·0 to 10·5, period 312 days. A long-period M-type variable, visible to the naked eye near maximum.

LEO MINOR. A faint group, about midway between Regulus in Leo and Merak in Ursa Major. It contains no star brighter than magnitude 4. The only object of interest is the M-type long-period variable R; magnitude 6·2 to 12·3, period 370 days.

VIRGO. In shape Virgo is rather like a roughly-drawn Y. The brightest star is of course Spica; others are Gamma (2·8), Epsilon

279

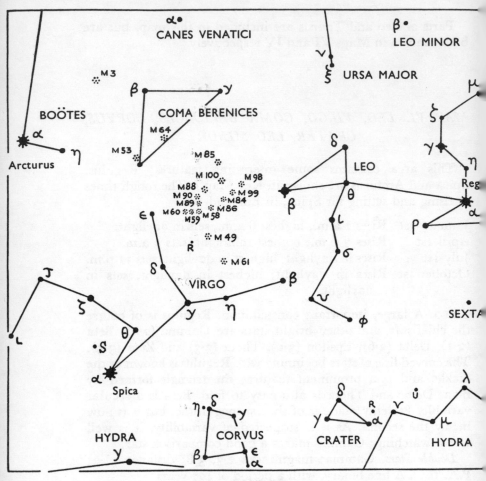

MAP VI

(2·9) and Zeta (3·4). The "bowl" of the Y, in the area enclosed by it and Beta Leonis, is very rich in faint galaxies, and is well worth sweeping.

Double Stars. Gamma: magnitudes 3·6, 3·7; distance about 5″. This is a magnificent binary, with a period of 172 years, and one of the best of all double stars for small telescopes.

Theta: magnitudes 4·0, 9·0; distance 7″; P.A. 343°. There is a 10th-magnitude star at a distance of 71″, making a useful test.

Variables. R: magnitude 5·9 to 12·0; period 145 days.

S: magnitude 5·6 to 12·3; period 372 days. Like R, a long-period variable of spectrum M.

COMA BERENICES and CANES VENATICI lie in the area enclosed by the Great Bear, Regulus, Beta Leonis and Arcturus. Coma contains no star brighter than magnitude 4½, but it is a rich area, and to the naked eye looks almost like a very scattered star-cluster, so that it is worth sweeping. Canes Venatici has one star, Alpha, of magnitude 2·9; it is a wide optical double; magnitudes 3·0, 5·6, distance 20″, P.A. 228°. Canes Venatici is so far north that it is circumpolar in England.

There are many clusters and nebulæ in these two constellations, which are shown on the map and are well worth looking for.

BOÖTES is shown in part, but is described with Map IX.

HYDRA is also partly shown, the brightest star being Nu (3·1). In this part of the constellation lies the interesting red N-type irregular variable U Hydræ, which has a magnitude range from 4·5 to 5·9.

CRATER. The brightest stars in this small group are Delta (3·8), Gamma (4·1) and Alpha (4·2), which form a triangle not far from Nu Hydræ. Not far from the reddish Alpha is the very red irregular variable R Crateris, with a magnitude range of from 8 to 9.

CORVUS. This is easy to find, as its four chief stars are of about the third magnitude (Gamma, 2·6; Beta, 2·7; Delta and Epsilon, each 3·0) and form a quadrilateral. To find it, pass a line from Arcturus midway between Spica and Gamma Virginis, the double star at the branch of the "Y". Delta Corvi is a double; magnitudes 3·1, 8·2; distance 24″; P.A. 212°.

Of the groups in this map, all are circumpolar apart from sections of Andromeda and Triangulum. Ursa Minor and Camelopardus are also shown, but are described with Map III.

CASSIOPEIA, shown in Key Map I, is one of the most interesting and conspicuous of the northern constellations. The Milky Way passes through it, and there are many rich telescopic fields. Of the chief stars, Alpha (Shedir) and Gamma are variable; the others are Beta (2·3), Delta (2·7) and Epsilon (3·4), which of course serve as excellent comparison stars.

Double Stars. Alpha: magnitudes 2·2 (var.) and 9·0; distance 63″; P.A. 280°. A wide optical double.

Eta: magnitudes 3·7, 7·3; distance 11″·2; P.A. 298°. Binary.

Iota: a fine triple. Magnitudes 4·2, 7·1, 8·0; distances 2″·3, 7″·5; P.A.s 251°, 113°.

Variables. Gamma: magnitude 1·7 to 3·4: irregular, and now classed as a "pseudo-nova". A most peculiar star, with a most unusual spectrum. Between 1950 and 1967 its magnitude averaged around 2·5. It is well worth watching.

Alpha. This was long classed as a variable: recently doubts have been cast on the reality of the fluctuations, but my own rough observations between 1936 and the present time indicate that the magnitude fluctuates irregularly between 2·1 and 2·5. Also worth watching.

R: magnitude 5·3 to 12·4; period 432 days. Spectrum M. *Clusters and Nebulæ.* M.52: a fairly bright cluster. Alpha and Beta act as "guides" to it.

M.103; An open cluster close to Delta.

CEPHEUS. This is not too easy to identify. The chief stars are Alpha (2·4), Beta (3·1), Gamma (3·2), Zeta (3·3) and Eta (3·4). Gamma lies between Beta Cassiopeiæ and Polaris; the main part of the constellation between Cassiopeia and Vega. The triangle made up of Zeta, Delta and Epsilon is the most conspicuous feature. On the whole, Cepheus is rather a barren group.

Double Stars. Beta: magnitudes 3·3, 8·0; distance 14″; P.A. 250°.

Kappa: magnitudes 4·0, 8·0; distance 7″·5; P.A. 122°.
Variables. Delta: magnitude 3·5 to 4·4; period 5·37 days. The prototype Cepheid.

Mu: magnitude 3·6 to 5·1; irregular. Sir William Herschel's "garnet star". It is of type M, and is probably the reddest of the naked-eye stars; a splendid object in a low power.

T: magnitude 5·5 to 9·6; period 391 days. Spectrum M.

V: magnitude 6·2 to 7·1; period 360 days. This lies between Gamma Cephei and Polaris.

LACERTA is a small constellation near Cepheus. It contains no star brighter than the 4th magnitude, and no objects of special interest.

PERSEUS. A grand constellation. It lies between Cassiopeia and Aldebaran; the chief star, Alpha (1·8) can be found by a line drawn from Gamma Cassiopeiæ through Delta Cassiopeiæ and prolonged. The other leading stars are Beta (Algol) (variable; 2·1 at maximum), Zeta (2·8), Epsilon and Gamma (2·9), Delta (3·0) and Rho (variable; 3·2 at maximum). The Milky Way is particularly rich in Perseus.

Double Stars. Zeta: magnitudes 2·8, 9·4; distance 12″·5; P.A. 208°. The chief component is a very luminous B1-type super-giant.

Eta: magnitudes 4·0, 8·5; distance 28″·4: P.A. 300°. The primary is yellow, the companion bluish.

Epsilon: magnitudes 2·9, 8·3; distance 9″; P.A. 009°.

Variable Stars. Beta (Algol); magnitude 2·1 to 3·3. The prototype eclipsing binary, fully described in the text.

Rho: magnitude 3·2 to 3·8; an M4-type irregular. A suitable comparison star is Kappa, magnitude 4·00.

Clusters. M.34; a fine open cluster, roughly between Kappa Persei and Gamma Andromedæ, visible to the naked eye on a transparent night.

H.VI.33 and 34. The "Sword-Handle" clusters, described in the text. They are visible to the naked eye, and in my view are the most beautiful of all open clusters. Between them is a faint red star.

ANDROMEDA. This is a bright constellation, the leading stars being Beta (2·0), Alpha and Gamma (2·1) and Delta (3·2).

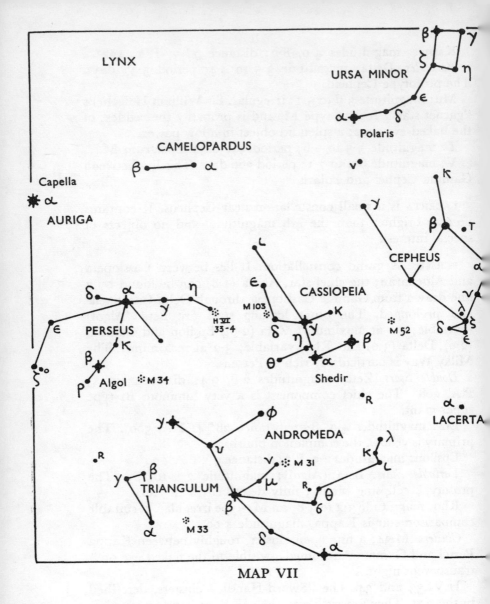

MAP VII

Alpha is included in the Square of Pegasus, and is also known as Delta Pegasi. It can be found by means of a line drawn from Epsilon Cassiopeiæ through Delta Cassiopeiæ, and prolonged.

Double Stars. Gamma; magnitudes 2·2, 5·0; distance 9"·8; P.A. 060°. A grand double, the components being yellow and blue. The small star is again double; magnitudes 5·4, 6·2; distance 0"·7; P.A. 109°.

Variable. R: magnitude 5·6 to 15; period 410 days. A long-period M-type variable, too faint at minimum for small apertures. It lies near Theta Andromedæ (4·4).

Galaxy. M.31; the Great Spiral, described in the text. It is visible to the naked eye as a misty patch close to Nu Andromedæ (4·4), but a telescope of large size is needed to show its structure.

TRIANGULUM. A fairly conspicuous little group near Andromeda, the leading stars being Beta (3·0) and Alpha (3·4).

Variable. R: magnitude 5·8 to 12; period 270 days. Spectrum M.

Galaxy. M.33. A large but rather faint and ill-defined object, roughly between Alpha Trianguli and Beta Andromedæ.

LYNX. One of the most barren of all constellations. It adjoins Camelopardus, and lies between Ursa Major and the Twins (Castor and Pollux). There are no bright stars or interesting telescopic objects worthy of mention here.

This is a very rich area, best seen in summer. Vega and Deneb are just circumpolar in England, and the approximate time of rising and setting for Altair are given below. It must be remembered that in all these "rising and setting" tables, allowance must be made for Summer Time.

January 1st   Rises 6 a.m., highest in daylight, sets 8 p.m.
April 1st   Rises midnight, highest 7 a.m., sets in daylight.
July 1st   Rises in daylight, highest 1 a.m., sets in daylight.
October 1st   Rises in daylight, highest 7 p.m., sets 2 a.m.

Vega and Deneb are shown on the first key map. Vega is almost overhead at midnight near midsummer, and can be recognized by its brilliance and by its bluish colour, which differs strongly from the yellowish hue of Capella, which occupies the overhead position at times during the winter.

LYRA. Though Lyra is a small constellation, and Vega is the only star above the third magnitude, it is remarkably rich in telescopic and other interesting objects. After Vega, the leading stars are Gamma (3·2) and the eclipsing binary Beta. The quadrilateral made up of Beta, Gamma, Delta and Zeta is easily recognized.

Double Stars. Epsilon. The famous double-double, described in the text. The two main components can be split with the naked eye; magnitudes 4·5 and 4·7; distance 208″. Epsilon[1]; magnitudes 4·6, 6·3; distance 2″·8; P.A. 001°. Epsilon[2]: magnitudes 4·9, 5·2; distance 2″·2; P.A. 099°. Of the two, Epsilon[1] is the easier to divide, but both pairs are well visible in a 3-inch O.G.

Zeta: magnitudes 4·3, 5·9; distance 44″; P.A. 150°. A wide, easy double.

Eta: magnitudes 4·5, 8·0; distance 28″; P.A. 083°.

Vega: has a companion, magnitude 10·5, at a distance of 56″ and a P.A. of 169°. This is an optical pair, not a binary system. The faintness of the companion makes it a convenient test object.

Variables. Beta: magnitude 3·4 to 4·4, period 12·9 days.

Eclipsing binary, described in the text. Gamma is a good comparison star; others are Zeta (4·1) and Kappa (4·3). It may be added here that the magnitude of 4·1 for Zeta as seen with the naked eye is the result of the combined light of the 4·3 and 5·9 magnitude components.

R: magnitude 4·0 to 4·7; a red M-type irregular variable.

Nebula. M.57. Planetary. The Ring Nebula, described in the text. It can be seen with a small aperture, but the central star is extremely difficult even with large instruments. The object is easy to find, as it lies directly between Beta and Gamma Lyræ.

CYGNUS. The Swan, but also, and perhaps more appropriately, known as the Northern Cross. It is a superb constellation, in a rich part of the Milky Way. The chief star is Deneb; other bright stars are Gamma (2·2), Epsilon (2·5), Delta (2·9), Beta (3·1) and Zeta (3·2). It is worth remembering that Beta, the faintest of the stars forming the Cross, lies roughly between Vega and Altair.

Double Stars. Beta (Albireo): magnitudes 3·1, 5·1; distance 34″·6; P.A. 055°. Yellow primary, green companion. I regard this as the loveliest double in the sky, and it is a superb object in any small telescope.

Delta: magnitudes 3·0, 6·5; distance 2″; P.A. 240°. A well-known test. Binary, with a period of 321 years.

61: magnitudes 5·6, 6·3; distance 28″; P.A. 142°. The celebrated star which was the first to have its distance measured.

Zeta: magnitudes 3·3, 7·9; distance 2″·3. Binary; period 500 years.

Variables. Chi: magnitude 4 to 14; period 409 days. A good comparison star when Chi is near maximum is its companion Eta (4·03).

W: magnitude 5·0 to 7·0; period 133 days. An M-type variable. It lies close to Lambda (4·5).

X: magnitude 6·0 to 7·0; period 16·4 days. A Cepheid, lying close to Rho (4·2).

Albireo has been suspected of very long period or secular variability, but this remains unconfirmed.

Nebulæ and Clusters. There are many nebular objects in Cygnus. One of the most striking is M.39, near Rho, a good open cluster and a fine sight in a small telescope.

VULPECULA is a small constellation near Cygnus. It contains no star brighter than magnitude 4½. The most interesting object is M.27, the Dumb-bell Nebula, a planetary; it is dim, but is well worth looking at, even though a telescope of some size is needed to show it properly. It lies not far from Gamma Sagittæ. Vulpecula, the Fox, was once known as Vulpecula et Anser, the Fox and Goose; but nowadays the goose seems to have been discarded—possibly the fox has eaten it!

DELPHINUS. A beautifully compact little group, very easy to recognize. Vega and Beta Cygni point to it. The brightest star is Beta (3·7). The most interesting object is the double star Gamma; magnitudes 4·5, 5·5; distance 10″·5; P.A. 270°. The primary is yellow, the companion green.

SAGITTA. Another compact group; the brightest stars are Gamma (3·7) and Delta (3·8). It lies between Altair and Beta Cygni.

EQUULEUS. The chief star of this little constellation is Alpha (4·1). Delta is an excessively close double, and a rapid binary.

PEGASUS. Most of this constellation, including the Square, is shown on Map X. The chief star in the present map is Epsilon (2·3), which is suspected of variability. Close to it lies the bright globular cluster M.15, a fine sight in a moderate telescope.

AQUARIUS also lies mainly in Map X. On the present map are Beta (2·9), Alpha (3·0), and two nebulous objects; the fine globular M.2, which I find fully resolvable with my 12½-in. reflector and which lies between Beta Aquarii and Epsilon Pegasi, and the beautiful planetary H.IV.1, which lies in the same low-power field as the orange star Nu Aquarii (4·5). Aquarius is a Zodiacal constellation.

AQUILA. The chief star, Altair, is of the 1st magnitude, and is easy to recognize because it has a brightish star to either side, Beta and Gamma. As well as Altair, the constellation includes Gamma (2·7), Zeta (3·0), Theta (3·1), Delta and Lambda (3·4) and Beta (3·9). The line below Altair, made up of Theta, Eta and Delta, is very easy to identify.

Variables. Eta: magnitude 3·7 to 4·5, period 7·2 days. A typical Cepheid.

R: magnitude 5·7 to 12; period 310 days. Spectrum M.

SERPENS. This constellation is divided into parts, Cauda (the body) and Caput (the head), separated by Ophiuchus. Caput is shown in Map IX. The brightest star in Cauda is Eta (3·2); the most interesting object is the fine double Theta, magnitudes 4·5 and 4·5, distance 22″, P.A. 103°. This is a splendid object, and is easy to recognize, as it lies in a rather isolated position not far from Delta Aquilæ.

SCUTUM. Though containing no star brighter than the fourth magnitude, Scutum lies in a rich part of the Milky Way, and shows some fine fields. There are several clusters. One of these is the "Wild Duck", M.11, one of the most beautiful open clusters in the sky, and shaped like a fan; it lies near Lambda Aquilæ. M.26, close to Delta Scuti (4·7) is another good open cluster. It is well worth while to sweep this whole region with a low power.

SAGITTARIUS. This is a large and bright constellation, but is always very low in New York, and cannot be seen to advantage; part of it never rises at all. The chief stars are Epsilon (1·8), Sigma (2·1), Zeta (2·6), Delta (2·7), Lambda (2·8), Pi (2·9), Gamma (3·0), Eta (3·2) and Tau (3·3). Deneb, Altair and Sagittarius lie almost in a straight line, with Altair in the middle; this is probably the easiest way to find Sagittarius. It can be quite conspicuous on summer evenings. Adjoining Sagittarius, but well south, is the little constellation CORONA AUSTRALIS (the Southern Crown).

Clusters and Nebulæ. M.17; the Omega or Horseshoe Nebula, near Gamma Scuti; a fine object in a moderate telescope.

M.8; the Lagoon Nebula, an easy object near Mu Sagittarii.

M.22; a bright globular between Sigma and Mu, not far from Lambda.

CAPRICORNUS. Like Sagittarius, Capricornus is in the Zodiac. It is rather a barren group; the chief stars are Delta and Beta (each 2·9).

Double Stars. Alpha: magnitudes 3·7, 4·3; distance 376″. This is a naked-eye double, and is easy to find, as the line of

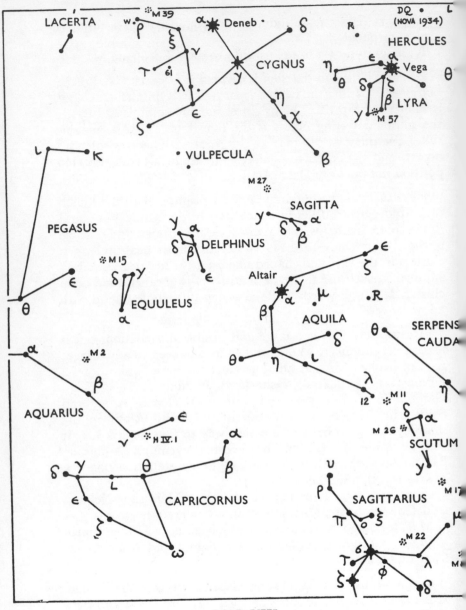

MAP VIII

stars made up of Gamma Aquilæ, Altair and Beta Aquilæ points to it. The fainter component is again double; 3·7, 11; distance 7″; P.A. 158°, and the smaller component of this pair is again double, though a very difficult object.

Beta; a very wide double. Magnitudes 3·1, 6; distance 205″, P.A. 290°. The fainter component is again double; distance 1″·3, P.A. 103°, but the companion is rather faint (10·6) and is thus rather difficult in small apertures.

HERCULES. A small part of Hercules appears in Map VIII, but most of the constellation lies in Map IX. The site of the 1934 nova, DQ Herculis, is marked. This is now a difficult object, and has been found to be a spectroscopic binary. It is described in the text.

MAP IX. BOÖTES, CORONA BOREALIS, HERCULES, SERPENS, OPHIUCHUS, LIBRA, SCORPIO

These are mainly summer groups, though the northernmost parts of Hercules and Boötes are circumpolar in England. Rough times of rising and setting for Antares, in Scorpio, are as follows:

January 1st   Rises 5 a.m., highest in daylight, sets in daylight.
April 1st   Rises 11 p.m., highest 3 a.m., sets in daylight.
July 1st   Rises in daylight, highest in daylight, sets 1 a.m.
October 1st Rises in daylight, highest in daylight, sets in daylight.

Arcturus in Boötes is easily recognized, and is shown on Key Map I. Corona is also most conspicuous, and can hardly be mistaken. The other groups are less easy to identify, as they are of large area but contain few bright stars. Scorpio is of course an exception, but the most brilliant part of the constellation is always very low in New York.

BOÖTES. The chief star is Arcturus; others are Epsilon (2·4), Eta (2·7), Gamma (3·0) and Delta and Beta (3·5). Arcturus is of type K, and is distinctly orange.

Double Stars. Epsilon: magnitudes 2·5, 5·3; distance 3″, P.A. 340°. The primary is yellowish, the companion bluish.

Zeta: magnitudes 4·6, 4·7; distance about 1″·3, P.A. 135°. This is close and rather difficult. Binary, period 123 years.

Xi: magnitudes 4·8, 6·9; distance 7″·0, P.A. 344°. Binary, period 152 years.

Delta: magnitudes 3·5, 7·8; distance 105″, P.A. 080°. A very easy object in a small telescope.

Variables. W and R, which lie close to Epsilon. R varies from magnitude 6 to 13 in 225 days; W from 5·2 to 6, irregular.

CORONA BOREALIS. This beautiful little constellation can hardly be mistaken, and it really does look rather like a "crown". The chief stars are Alpha (2·2) and Beta (3·7). Despite its small size, Corona is rich in interesting objects.

Double Stars. Eta: magnitudes 5·7, 5·9; distance 1″, P.A. varies rather quickly, as the star is a binary with a period of 42 years. It is rather close, and is thus not an easy object.

Zeta: magnitudes 4·0, 4·9; distance 6″·3, P.A. 303°. A fine double.

Variables. T: the peculiar nova-like variable. Usually it fluctuates between magnitudes 9 and 10, but it rose to 2 in 1866 and to 3 in 1946. It is well worth watching, as a fresh outburst may occur at any moment.

R: magnitude 5·6 to 12·5. The well-known irregular variable, described in the text.

S: magnitude 6 to 12, period 361 days. Spectrum M.

HERCULES. A very large but rather barren constellation. It occupies the area between Vega and Corona Borealis. The chief stars are Beta and Zeta (2·8), Alpha (variable), Pi and Delta (3·1), Mu (3·4) and Eta (3·5).

Double Stars. Zeta: magnitudes 3·1, 5·6; distance about 1″. P.A. alters fairly quickly, as the star is a binary with a period of 34 years.

Delta: magnitudes 3·2, 7·5; distance 11″, P.A. 208°.

Alpha: magnitudes 3 (variable), 5·4; distance 4″·6, P.A. 110°. The brighter star is an M5-type giant, reddish; the companion green.

Variables. Alpha. One of the Betelgeux-type irregulars. It fluctuates between magnitudes 3 and 3½, and over about twenty years I have found no semblance of a period. The best comparison stars for it are Kappa Ophiuchi (3·18), Gamma Herculis (3·79) and Delta Herculis (3·14).

g: magnitude 4·6 to 6·0. An M-type irregular, near Sigma (4·2).

S: magnitude 6 to 12·5, period 300 days. Spectrum M. It lies between Alpha Herculis and Beta Serpentis.

Clusters. M.13. The famous globular; it lies between Zeta and Eta, and can just be seen with the naked eye under good conditions. It is very easy to find with a telescope, and in a moderate aperture is a glorious sight.

M.92: another globular, between Iota and Eta. It is not unlike M.13, but is far less prominent.

Σ5: a small bright planetary nebula, in the triangle formed by Beta, Delta and Epsilon Herculis. It is said to have a bluish hue, though to me it always looks white.

OPHIUCHUS. This constellation lies between Vega and Antares. It contains some fairly bright stars: Alpha (2·1), Eta (2·5), Zeta (2·6), Delta (2·7), Beta (2·8), Kappa and Epsilon (3·2), and Mu and Nu (3·3), but it is not easy to identify at first sight, and it is relatively barren of interesting objects. There is a bright globular cluster, M.19, near Theta, and roughly between Theta and Antares; but it is always very low in New York. Ophiuchus is not classed as a Zodiacal constellation, but it does enter the Zodiac in the region between Scorpio and Sagittarius.

LIBRA. Zodiacal, but a very dull constellation. The chief stars are Beta (2·6), Alpha (2·8) and Sigma (3·3); Sigma is also included in Scorpio, as Gamma Scorpionis. There are few interesting objects apart from the Algol-type eclipsing binary Delta Libræ, which has a magnitude range of 4·8 to 6·2 and a period of 2·3 days. Beta Libræ is a B8-type star, and is said to be the nearest approach to a normal "green" star. It certainly may have a slightly greenish tinge, though the colour is so elusive that many people will fail to detect it. Of course, some double stars have green components, and Nova DQ Herculis was also green at one stage in its career.

SCORPIO. A splendid Zodiacal group, but never well seen in New York as it is always low down; from California

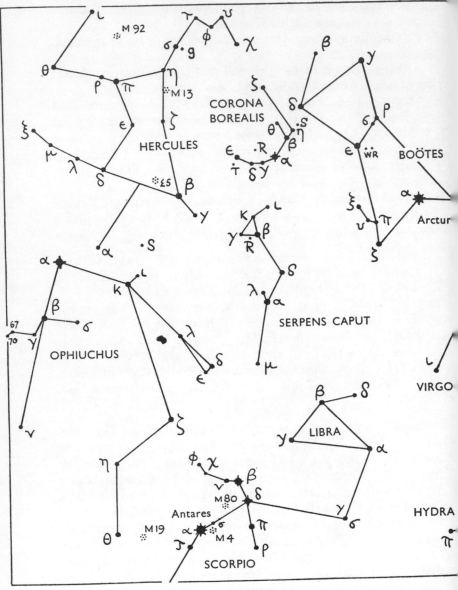

MAP IX

and Arizona, however, it is most impressive. The chief stars, apart from Antares, are Lambda (1·6), Theta (1·9), Epsilon and Delta (2·3), Kappa (2·4), Beta (2·6), Upsilon (2·7), Sigma and Tau (2·8), Pi (2·9), Iota[1] and Mu (3·0), G (3·2) and Eta (3·3), but Lambda, Upsilon, Kappa, Iota, Theta and Eta are invisible in England. Regulus and Antares are on roughly opposite sides of Arcturus, with Arcturus in the middle, which is of help in identifying Scorpio; Antares is also distinguished by its ruddiness, and by the fact that, like Altair, it has a fairly bright star to either side of it—in this case Tau and Sigma Scorpionis.

Double Stars. Antares has a companion of magnitude 5·1; distance 3″, P.A. 275°. The primary is of course red; the companion is green. It is a fine object.

Nu: magnitude 4·3, 6·5; distance 41″; P.A. 335°. A wide, easy double. Each component is again double, but very close and difficult.

Beta: magnitudes 2·8, 5·0; distance 1″. There is a third star, magnitude 4·9, at 14″.

Clusters. M.80. A splendid globular, lying roughly between Antares and Beta.

M.4. An open cluster. The stars in it are not brilliant, but the object is not hard to find, as it lies close to Antares.

SERPENS. The chief star in Caput is Alpha (2·6). R is an M-type variable; 5·5-13·4, 357 days. M.5, a bright globular, lies near Alpha.

MAP X. PEGASUS, ANDROMEDA, PISCES, TRIANGULUM, ARIES, CETUS, AQUARIUS, SCULPTOR, PISCIS AUSTRALIS

The chief group in this map is the Square of Pegasus, which in my view is much more difficult to identify than might be supposed, since most people expect it to be smaller and brighter than it really is. The best way to find it is by means of Cassiopeia, since Gamma and Alpha Cassiopeiæ point directly to it. The line from Merak and Dubhe through Polaris will also reach the Square if prolonged far enough across the sky. Very rough risings and settings are as follows:

January 1st Rises in daylight, highest in daylight, sets at midnight.
April 1st   Rises 2 a.m., highest in daylight, sets in daylight.
July 1st   Rises in daylight, highest 5 a.m., sets in daylight.
October 1st Rises in daylight, highest 11 p.m., sets 7.a.m.

It is therefore at its best during the autumn. As is shown on Map VII, one of the stars of the Square is generally included in the neighbouring constellation of Andromeda (Alpha Andromedæ=Delta Pegasi). Andromeda and Triangulum are described with Map VII.

PEGASUS. An important constellation, but not so conspicuous as is generally supposed. Alpha Andromedæ (2·1) is in the Square. The other chief stars of Pegasus are Epsilon (2·3), which is shown on Map VIII, Alpha (2·5), Beta (variable), Gamma (2·8), Eta (2·9) and Zeta (3·4). It is rather instructive to count the number of stars inside the Square visible with the naked eye; there are not very many of them.

Double Star. Xi: magnitude 4·0, 12; distance 12″, P.A. 108°. A difficult double, owing to the faintness of the companion. It is a binary, with a period of about a century and a half.

Variable. Beta: magnitude 2·4 to 2·7. An M-type irregular. Suitable comparison stars are Alpha (2·50) and Gamma (2·84). There is a very rough period of about 35 days.

ARIES. Celebrated as being the First Constellation of the Zodiac. It is not, however, very conspicuous. It lies between

Aldebaran and the Square of Pegasus, and has two fairly bright stars, Alpha (2·0) and Beta (2·7).

Double Star. Gamma: magnitudes 4·7, 4·8; distance 8″·5, P.A. 000°. A fine, easy double, very well seen with a small telescope.

PISCES. The last constellation of the Zodiac, though owing to the precession of the equinoxes it now contains the First Point of Aries. It is large but faint, the brightest star being Eta (3·7). Pisces can be identified by the long line of rather faint stars running below the Square of Pegasus.

Double Stars. Alpha: magnitudes 4·3, 5·3; distance 1″·9, P.A. 292°.

Zeta: magnitudes 4·2, 5·3; distance 24″, P.A. 060°.

CETUS. Part of this large constellation is shown in Map IV, and the chief star (Beta) in the key map. Beta can be found by means of the Square of Pegasus, since Alpha Andromedæ and Gamma Pegasi point towards it. Its proper name, Diphda, is often used, and it is an orange star suspected of variability. Not far from it is the M-type irregular variable T, which is reddish, and has a magnitude range of from 5 to 7. Mira (Omicron), shown here, is described with Map IV.

SCULPTOR (a merciful abbreviation of the old name "Apparatus Sculptoris"). A very obscure constellation near Diphda. It contains no star as bright as the fourth magnitude, and no objects of interest to the amateur.

AQUARIUS. Part of this Zodiacal constellation is shown in Map VIII, but most of it lies in the present map. The chief stars are Beta (2·9) and Alpha (3·0); (Map VIII); Delta (3·3) and Zeta (3·7). There is a striking group of orange stars centred round Chi (5·1); these are easy to identify, and make pleasing telescopic objects under a low power.

Double Star. Zeta: magnitudes 4·4, 4·6; distance 1″·9, P.A. 256°. A fine binary, with a period of 360 years.

Variable: R, magnitude 6 to 11, period 380 days. An M-type long-period variable, not far from the star Omega[2] (4·6).

PISCIS AUSTRINUS. This small group is also termed Piscis Australis. It contains Fomalhaut, of the 1st magnitude, but no

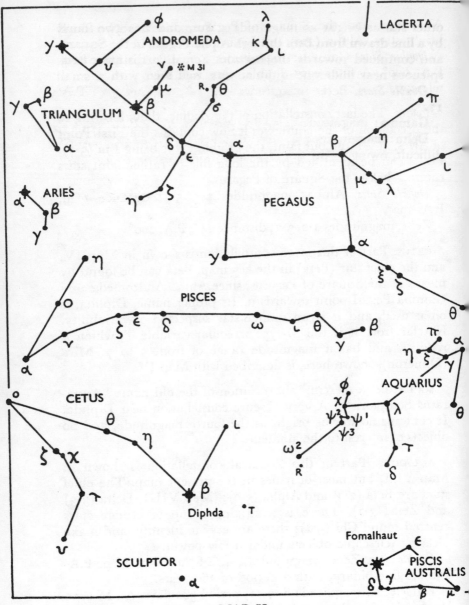

MAP X

other star as bright as magnitude 4. Fomalhaut can be found by a line drawn from Beta through Alpha Pegasi, in the Square, and continued towards the horizon. Fomalhaut is quite conspicuous near midnight in the autumn months.

Double Stars. Beta: magnitudes 4·4, 7·8; distance 30″, P.A. 172°.

Gamma: magnitudes 4·5, 8·5; distance 4″·3, P.A. 262°.

Delta: magnitudes 4·3, 10·6; distance 5″, P.A. 240°. Rather difficult, owing to the faintness of the companion.

MAPS XI, XII, XIII and XIV.
THE FAR SOUTHERN CONSTELLATIONS

These maps are drawn to a smaller scale, and are described only briefly. This is for the excellent reason that I have never been south of the equator, and have never seen the stars which remain below the horizon in Britain and the United States, so that I cannot be expected to act as a guide to others. There is no doubt that the stars in the far south are of great interest, and there are many superb objects, such as the Magellanic Clouds.

It is probably true to say that the great majority of astronomical books cater for readers in the northern hemisphere, and that in consequence the Australasian student feels rather neglected. For instance, it is often said that the Sun is due south at noon. It is—to Britons; but to Australians the Sun's direction at noon is due north. The vernal or "spring" equinox occurs in March; but March is not spring in Australia or New Zealand.

Things are even worse with regard to the star maps, since from the English point of view everything is upside-down. Orion's sword, including the great gaseous nebula M.42, hangs upward from the Belt; Rigel is in the top corner of Orion, Betelgeux in the lower; and so on. Unfortunately there is no solution to this difficulty, short of re-lettering the whole series of maps. Those in Australia or New Zealand should however have no difficulty in using the maps I have given here—all they have to do is to turn the book round.

The whole aspect of the night-sky is different. The Southern Cross is circumpolar for much of Australia; the grand constellation Scorpio, never well seen from Britain, passes near the zenith; Canopus shines with a brilliancy rivalling that of Sirius. It is true that there are no bright stars very close to the South Celestial Pole, but once certain southern groups such as Centaurus and Crux Australis have been recognized—which should be done easily enough—the beginner should have no difficulty in finding his way about. Fortunately, *Norton's Star Atlas*, which I regard as quite indispensable for the night-sky student, is complete for the whole heavens.

MAP XI

This includes the southern part of Eridanus, with the brilliant Achernar, as well as Theta (Acamar) which was recorded of the first magnitude in Ptolemy's time, but has now faded to the third.

HOROLOGIUM is an obscure group; chief star Alpha (3·8).

CÆLUM, partly visible in New York, has already been mentioned.

RETICULUM is a compact little group; chief star Alpha (3·3).

DORADO has as its chief star Alpha (3·5); then comes Beta (3·8). The Nubecula Major lies mainly in Dorado, but extends into Mensa.

MENSA is otherwise an extremely obscure group, with no star as bright as the 5th magnitude.

HYDRUS has a few brightish stars: Beta and Alpha (each 2·8) and Gamma (3·3). Alpha Hydri is close to Achernar. A small part of the Nubecula Minor spreads into Hydrus, though most of it lies in Tucana.

CHAMÆLEON. Chief stars, Alpha and Gamma (4·1). Delta is a double: magnitudes 6·1, 6·4; distance 0"·5, P.A. 070°.

OCTANS has no star brighter than Nu (3·7), but contains the southern celestial pole. The nearest naked-eye star to the pole is Sigma (5·5).

APUS. A compact group; chief star Alpha (3·8). Theta is an M-type irregular variable, with a magnitude range of 5·0 to 6·6.

MUSCA AUSTRALIS has as its chief stars Alpha (2·7) and Beta (3·1). Beta is double: magnitudes 3·7, 4·0; distance 1"·3, P.A. 006°.

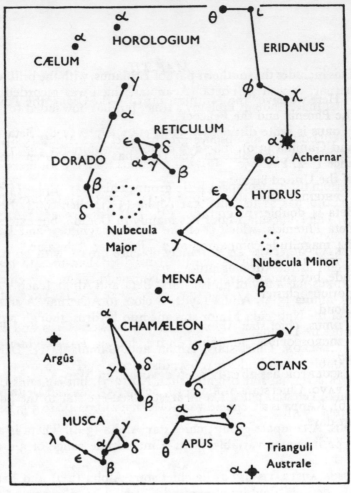

MAP XI

MAP XII

Here we have "the Southern Birds"—the Crane, the Toucan, the Phœnix and the Peacock.

GRUS is quite distinctive; chief stars Alpha (1·8), Beta (2·1) and Gamma (3·0). Theta is double: magnitudes 4·5, 7; distance 1″·4, P.A. 050°. It is conspicuous from the southern parts of the United States.

PHŒNIX. Chief stars Alpha (2·4), Beta (3·3), Gamma (3·4). Beta is double: magnitudes 4·0, 4·1; distance 2″, P.A. 355°. Zeta Phœnicis, which is of the fourth magnitude and has an 8·4 magnitude companion at 7″, lies near Achernar.

TUCANA has only Alpha (2·9) brighter than the 4th magnitude, but contains nearly all the Nubecula Minor, as well as the glorious cluster 47 Tucanæ, very close to the edge of the Cloud.

INDUS, chief star Alpha (3·1), lies between Grus and Pavo.

MICROSCOPIUM, adjoining Indus, has no star brighter than 4·7.

SCULPTOR has already been described in Map X.

PAVO. Chief stars Alpha (2·0), Beta (3·5), Eta and Delta (each 3.6). Kappa is a Cepheid variable: magnitude 4·0 to 5·5, period 9·1 days.

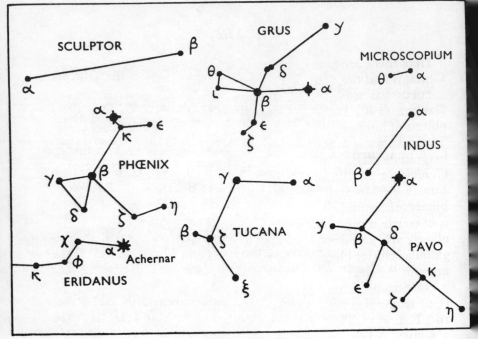

MAP XII

MAP XIII

Here there are many bright stars, including Alpha and Beta Centauri and the Southern Cross.

LUPUS has some brightish stars; Alpha (2·2), Beta (2·7), Gamma (2·8), Delta (3·2) and Zeta and Eta (3·4), but few objects of interest.

CENTAURUS is a brilliant group. In addition to its two first-magnitude stars, Alpha and Beta (Agena), it has Theta (2·0), Gamma (2·2), Eta and Epsilon (2·3), Zeta and Delta (2·9), Iota (2·8), Mu and Lambda (3·1) and Nu (3·4). Gamma is a binary: magnitude 3, 3; distance 1″·5, period 80 years. Alpha is of course a superb double: magnitudes 0·0, 1·4; a very easy object; also a binary, with a period of 80 years. Here too is the globular cluster Omega Centauri, from all accounts by far the finest in the sky.

CIRCINUS lies close to Alpha Centauri. The chief star, Alpha (3·2) is double; the companion is of magnitude 8·8, distance 16″, P.A. 235°. The primary is yellow, the companion reddish.

NORMA. A faint group between Lupus and Ara. There is no star as bright as the 4th magnitude.

TRIANGULUM AUSTRALE. This is a bright little group, the chief stars being Alpha (1·9), Beta and Gamma (each 2·9).

CRUX AUSTRALIS. The famous Southern Cross. Acrux and Beta are of the 1st magnitude; then follow Gamma (1·7), an M-type Red Giant, and Delta (2·8). Acrux is double: magnitude 1·4, 1·9; distance 4″·7, P.A. 119°. Surrounding Kappa Crucis, near Beta, is a magnificent open cluster. •

ARA. Chief stars Beta and Alpha (2·9), Zeta (3·2) and Gamma (3·3). Alpha and Beta Centauri point one way towards Ara and the other way towards Crux Australis. Gamma has a 10th-magnitude companion at 18″.

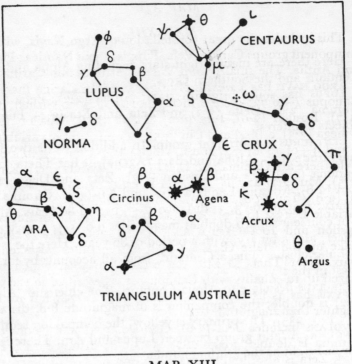

MAP XIII

MAP XIV

This includes the great constellation of Argo Navis, with its component groups Carina, Vela, Puppis; Pyxis Nautica; Pictor, and Antlia. The small group Volans intrudes into Carina.

ARGO NAVIS has a wealth of bright stars. In Carina there are Canopus (—0·73; second in brightness only to Sirius), Beta (1·7), Epsilon (2·0), Iota (2·2), Theta (2·7), Upsilon (2·9), Omega (3·3), a (3·4) and Chi (3·5), as well as three variables— p and q—of about 3·3, and l, which varies from 3·4 to 4·8. In Vela there are Gamma and Delta (1·9), Lambda (2·2), Kappa (2·5), Mu (2·7) and N (3·2); in Puppis, Zeta (2·2), Rho and Pi (2·8), Tau (3·0), Nu (3·2), Sigma and Xi (3·3) and the variable L^2 (3·4 to 6·2). Part of Puppis is visible in England. Epsilon and Iota Carinæ and Kappa and Delta Velorum make up the "false Cross", which is sometimes mistaken for Crux. Near Theta is Eta, the erratic variable, which once blazed up to equality with Canopus. It is described in the text.

PYXIS has as its chief star Alpha (3·7). ANTLIA has no star brighter than magnitude 4·4.

VOLANS includes Beta (3·6), Gamma (3·7) and Zeta (3·9). Gamma is double; magnitudes 3·9, 5·8; distance 14″, P.A. 300°. Zeta is also double: magnitudes 3·9, 9·0; distance 17″; P.A. 116°.

PICTOR, between Canopus and Dorado (Map XI) has two fairly conspicuous stars, Alpha (3·3) and Beta (3·9).

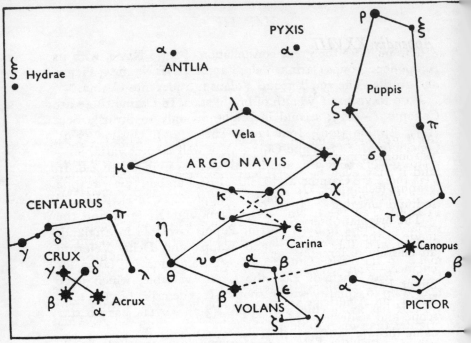

MAP XIV

308

Appendix XXVIII

RADIO ASTRONOMY

As I POINTED OUT in Chapter Thirteen, I am not a radio astronomer; moreover this book is concerned with visual work, and it may even be that I have said too little about photographic techniques. On the other hand, radio astronomy is now so important that it cannot be ignored.

The original observations of radio waves from the sky were made in the early 1930s by Karl Jansky, who was working on an entirely different programme—he was a communications engineer—and who never followed up his discoveries as he might have been expected to do. A few years later Grote Reber, an American amateur, built the first true radio telescope and undertook researches of pioneer importance. Then came the war, during which radio and radar techniques were developed to a remarkable degree. After the end of hostilities, it seemed as though researches of this kind, including radio astronomy, would remain solely in the hands of professional scientists.

As far as really fundamental advances are concerned, this is of course true. The cost of equipment even remotely comparable with that at a professional establishment would be far beyond the means of the wealthiest amateur, and to set up an installation of such a kind would be rather pointless in any case. On the other hand, it has now been shown that the serious amateur is capable of useful work if he is prepared to spend time and a certain amount of money in constructing adequate equipment. Oddly enough, the main cost lies in the recording devices rather than in the aerials themselves.

In its simplest form, a radio telescope consists of an aerial together with a radio receiver and a recording device. A "radiometer" of this sort may be used for the measurement of daily changes of intensity of, for example, the radio emissions from the Sun, and it may be designed so as to cover several frequencies at the same time. If sufficient space is available, radiometers or aerial systems may be combined to make an "interferometer". Steerable radio telescopes may also be constructed. It is obvious that the would-be astronomer must have a thorough working knowledge of electronics; in fact he must be an electronics man first and an astronomer afterwards.

Radio telescopes are of many designs. The "dish", of which the greatest example is the 250-foot paraboloid at Jodrell Bank, is the best-known, but other instruments do not look like telescopes in any sense of the word, and give the general impression of a haphazard collection of poles and wires. I do not propose to go into details here; this is best left to a radio astronomer, and I have listed some useful books on Page 317. Meanwhile, it may be worth giving a few notes about the types of radio sources to be found in the sky.

First, of course, there is the Sun, whose radio emissions may be studied with amateur equipment. Large-scale disturbances may be recorded from the active Sun; with the quiet Sun, the slowly-varying component is observable, and flares produce very marked effects. There are also more specialised investigations. Each year the Sun passes close to the famous Crab Nebula in Taurus, the supernova wreck which is itself a powerful radio source. The Nebula is occulted by the solar corona, and the various effects produced are of great significance. Amateurs can do valuable work here, beginning their observations in mid-May and continuing until mid-July.

Emissions from Jupiter come into a different category. As we have seen, they were formerly thought to be due to Jovian thunderstorm activity or something of the kind. This was questioned by F. W. Hyde, the noted British amateur who has established a fully-fledged radio astronomy observatory at Clacton; according to Hyde, the emissions are related to solar activity. Since 1963 a major programme has been going on, undertaken by Hyde in collaboration with professional scientists at Florida State University in the U.S.A. under C. H. Barrow, and it is starting to look as though his theory is correct, though the question has not yet been definitely settled. Here we have a first-class example of co-operation between amateur and professional radio astronomers, though it is only fair to add that so far, at least, Hyde's observatory is in a class of its own. (Visual observers of Jupiter are also taking part, since it is obviously essential to know just what is happening on the planet's surface.)

Most of the discrete radio sources inside our own Galaxy, outside the Solar System, appear to be the wrecks of old supernova. Several of them are easily available to the amateur worker, notably Cassiopeia A. Here we have a strong radio source, identified with some excessively faint nebulosity visible only with large instruments. The most interesting point, perhaps, is that the intensity of the Cassiopeia A radio emission is thought to be diminishing steadily each year, and measures made to a well-calibrated standard will be most useful if they are extended over a long period of time. Beyond the Galaxy lie the other radio sources, such as the powerful Cygnus A, and there are of course the quasars, which are undoubtedly the most enigmatical of all the objects in the sky.

Many research fields in radio astronomy are closed to the amateur, simply on the grounds of insufficient equipment, but other branches are open, and have not yet been exploited. There is immense scope here, though, of course, anyone who wants to build and use a radio telescope will have many problems to face—and will be wise to do some serious reading before starting out!

ASTRONOMICAL SOCIETIES

ANY SERIOUS amateur will be well advised to join a society. The advantages of doing so are obvious; he will be able to collaborate with others, and to exchange information and points of view. Incidentally, he will make many friends.

In Britain, the leading amateur organization is the British Astronomical Association (B.A.A.). The secretarial address, from which all information may be obtained, is 303 Bath Road, Hounslow West, Middlesex. No qualifications other than patience and enthusiasm are needed for entry. Monthly meetings are held at Burlington House; the *Journal* is published monthly, together with various other publications. There are specific sections which members may join if they wish, and the B.A.A.'s observational record is second to none.

For younger enthusiasts, there is the Junior Astronomical Society (44 Cedar Way, Basingstoke, Hampshire) which issues a quarterly periodical, *Hermes*, and holds London meetings. There are also provincial branches. The J.A.S. is affiliated to the B.A.A.

There are naturally many local astronomical societies, such as those of Liverpool, Bristol, Leeds and Norwich (to mention only four). In fact, few large cities are without them. A full list is published annually in the *Yearbook of Astronomy*. Of course, many professional organizations exist, pre-eminent among which is the Royal Astronomical Society, but to discuss these would be beyond our present scope.

In Ireland, there is the Irish Astronomical Society, with centres at Belfast, Armagh, Londonderry and Dublin. The chief secretarial address is 1 Garville Road, Dublin 6. The other secretarial addresses are 35 Ardenvohr Street, Belfast; Magee College, Londonderry, and The Planetarium, Armagh.

Amateur societies also flourish in many foreign countries, and some have their own observatories. Special mention should be made of the Association of Lunar and Planetary Observers

(Box AZ, University Park, New Mexico, U.S.A.) which publishes regularly *Strolling Astronomer*, edited by W. H. Haas. This Association is predominantly amateur. The Astronomical Society of the Pacific (675 Eighteenth Avenue, San Francisco 21, California, U.S.A.) and the American Association of Variable Star Observers (A.A.V.S.O.) contain many amateur as well as professional members. The A.A.V.S.O. also has a section devoted to solar work.

The Royal Astronomical Society of Canada, with several branches, also includes many amateurs; so does the Royal Astronomical Society of New Zealand, with its secretarial address at the Carter Observatory, Wellington.

Further details of Astronomical Societies in the United States are listed in the *Yearbook of Astronomy*, published annually by W. W. Norton & Company, Inc., New York.

BIBLIOGRAPHY

ASTRONOMICAL literature is so vast that it is quite out of the question to give an even approximately full book list. All I have tried to do is to give enough references for the beginner. Emphasis has been laid on popular and semi-popular works; those somewhat more technical are distinguished by an asterisk. I have not attempted to include periodicals, but mention must be made of *Sky and Telescope*, published by the Sky Publishing Corporation, Cambridge 38, Massachusetts, U.S.A. This appears monthly.

YEARBOOK

Yearbook of Astronomy. Annually.

GENERAL ASTRONOMY

BOK, B. J. *The Astronomer's Universe.*

CHISNALL, G., and FIELDER, G.* *Astronomy and Spaceflight.*

LOVELL, BERNARD, and LOVELL, JOYCE. *Discovering the Universe.*

MOORE, PATRICK. *Astronomy.*

MOORE, PATRICK. *The Amateur Astronomer's Glossary.*

MOORE, PATRICK. *The Sky at Night.*

MOORE, PATRICK (editor) . *Practical Amateur Astronomy.*

MUIRDEN, J. *Astronomy with Binoculars.*

PAYNE-GAPOSCHKIN, C. *Introduction to Astronomy.*

PICKERING, J. S. *1001 Questions Answered about Astronomy.*

RICHARDSON, R. S. *The Fascinating World of Astronomy.*

RUDAUX, L., and DE VAUCOULEURS, G. *Encyclopedia of Astronomy.*

SIDGWICK, J. B. *Observational Astronomy for Amateurs.*

STRUVE, O. *Elementary Astronomy.*

THE SUN

ABETTI, G. *Solar Research.
ABETTI, G. The Sun.
BAXTER, W. M. The Sun and the Amateur Astronomer.
ELLISON, M. A. The Sun and Its Influence.
KUIPER, G. P. (editor). *The Sun.
NEWTON, H. W. The Face of the Sun.
SEVERNY, A. The Sun.

THE MOON

BALDWIN, R. B. *The Measure of the Moon.
FIELDER, G. *Structure of the Moon's Surface.
FIELDER, G. Lunar Geology.
FIRSOFF, V. A. *Surface of the Moon.
MARKOV, A. V. (editor). *The Moon: a Russian View.
MOORE, PATRICK. Survey of the Moon.
MOORE, PATRICK, and CATTERMOLE, P. J. *Craters of the Moon.
MOORE, PATRICK. Map of the Moon (24-inch diameter).
SPURR, J. E. *Features of the Moon: *Lunar Catastrophic History: *The Shrunken Moon.

THE PLANETS

FIRSOFF, V. A. Exploring the Planets.
MOORE, PATRICK. The Planets.
UREY, H. C. *The Planets.
WHIPPLE, F. L. Earth, Moon and Planets.
(Mercury)
SANDNER, W. The Planet Mercury.
(Venus)
MOORE, PATRICK. The Planet Venus.
(Mars)
DE VAUCOULEURS, G. *Physics of the Planet Mars.
JACKSON, F. L., and MOORE, P. Life on Mars.
MOORE, PATRICK. Guide to Mars.

SLIPHER, E. C. *Mars: The Photographic Story.*
STRUGHOLD, H. **The Green and Red Planet.*
(Jupiter)
PEEK, B. M. *The Planet Jupiter.*
(Saturn)
ALEXANDER, A. F. O'D. *The Planet Saturn.*
(Uranus)
ALEXANDER, A. F. O'D. *The Planet Uranus.*
(Minor Bodies)
ROTH, G. D. *The System of Minor Planets.*
SANDNER, W. *Satellites of the Solar System.*

AURORAE

HARANG, L. **The Auroræ.*
STORMER, C. **The Polar Aurora.*

COMETS AND METEOROIDS

FEDYNSKY, V. *Meteors.*
LOVELL, A. C. B. **Meteor Astronomy.*
LYTTLETON, R. A. *Comets and their Origin.*
NININGER, H. H. *Out of the Sky.*
PORTER, J. G. *Comets and Meteor Streams.*
RICHTER, N. *The Nature of Comets.*

ECLIPSES

MITCHELL, S. A. *Eclipses of the Sun.*

RADIO ASTRONOMY

HANBURY BROWN, A., and LOVELL, A. C. D. *The Exploration of Space by Radio.*

HYDE, F. W. *Radio Astronomy for Amateurs.*
LOVELL, A. C. B., and CLEGG, T. **Radio Astronomy.*
PFEIFFER, J. *The Changing Universe.*
SMITH, G. *Radio Astronomy.*

MATHEMATICAL ASTRONOMY

DAVIDSON, M. *Elements of Mathematical Astronomy.*

STELLAR ASTRONOMY

ABETTI, G. *Nebulæ and Galaxies.*
BRADE, W. **Evolution of Stars and Galaxies.*
MOORE, PATRICK. *Guide to the Stars.*
PAYNE-GAPOSCHKIN, C. *Stars in the Making.*
(Variable Stars)
CAMPBELL, L., and JACCHIA, L. *The Story of Variable Stars.*
PAYNE-GAPOSCHKIN, C. **Variable Stars and Galactic Structure.*
(Galactic Studies)
ABETTI, G., and HACK, M. *Nebulæ and Galaxies.*
SHAPLEY, G. *Galaxies.*

COSMOLOGY, ETC.

BONNOR, W. *The Mystery of the Expanding Universe.*
FIRSOFF, V. A. **Facing the Universe.*
MCVITTIE, G. C. **Fact and Theory in Cosmology.*
MOORE, PATRICK. *The New Look of the Universe.*
SCHATZMAN, E. *The Origin and Evolution of the Universe.*
SMART, W. M. *Origin of the Earth.*

STAR ATLAS

NORTON, A. P. *Star Atlas and Telescopic Handbook.* (This book has run to many editions. I regard it as indispensable.)

INSTRUMENTS, ETC.

HOWARD, N. *Handbook of Telescope Making.*

INGALLS, A. (editor). **Amateur Telescope Making.* Scientific American; last volume issued in 1953.

MATTHEWSON, G. *Constructing an Astronomical Telescope.*

MICZAIKA, G., and SINTON, W. M. **Tools of the Astronomer.*

ROTH, G. D. *The Amateur Astronomer and his Telescope.*

TEXEREAU, J. *How to Make a Telescope.*

WRIGHT, H. *The Great Palomar Telescope.*

HISTORICAL

ABETTI, G. *History of Astronomy*

ARMITAGE, A. *John Kepler.*

DREYER, J. E. *History of Astronomy from Thales to Kepler.*

HOSKIN, M. A. *William Herschel and the Construction of the Heavens.*

LEY, W. *Watchers of the Skies.*

PANNEKOEK, A. *History of Astronomy.*

RONAN, C. A. *Changing Views of the Universe.*

SHAPLEY, H., and HOWARTH, H. E. *A Source Book of Astronomy.*

DE VAUCOULEURS, G. *Discovery of the Universe.*

WRIGHT, H. *Explorer of the Universe* (G. E. Hale).

MISCELLANEOUS

ALFVÉN, H. *Worlds-Antiworlds.*

BERRILL, N. J. *Worlds Apart.*

DE VAUCOULEURS, G. *Astronomical Photography.*

FIRSOFF, V. A. **Life Beyond the Earth.*

KING-HELE, D. *Observing Earth Satellites.*

MOORE, PATRICK. *Space in the Sixties.*

MOORE, P., and JACKSON, F. L. *Life in the Universe.*

SHAPLEY, H. *Of Stars and Men.*

SPENCER JONES, SIR H. *Life on Other Worlds.*

THACKERAY, A. D. **Astronomical Spectroscopy.*

INDEX

322

323